Essential Electronics Series

Introduction to Power Electronics

Denis Fewson
Senior Lecturer
School of Electronic Engineering
Middlesex University

D1761062

A member of the Hodder Headline Group
LONDON • SYDNEY • AUCKLAND
Co-published in the USA by
Oxford University Press, Inc., New York

First published in Great Britain 1998 by Arnold,
a member of the Hodder Headline Group,
338 Euston Road, London NW1 3BH
http://www.arnoldpublishers.com

Co-published in the United States of America by Oxford University Press, Inc.,
198 Madison Avenue, New York, NY 10016
Oxford is a registered trademark of Oxford University Press

British Library Cataloguing in Publication Data
A catalogue record for this book is available from the British Library

Library of Congress Cataloging-in-Publication Data
A catalog record for this book is available from the Library of Congress

ISBN 0 340 69143 3 (pb)

Publisher: Nicki Dennis
Production Editor: Liz Gooster
Production Controller: Sarah Kett
Cover Designer: Terry Griffiths

Typeset in 10½/13½ Times by J&L Composition Ltd, Filey, N. Yorks
Printed and bound in Great Britain by J W Arrowsmith Ltd, Bristol

Series Preface

In recent years there have been many changes in the structure of undergraduate courses in engineering and the process is continuing. With the advent of modularization, semesterization and the move towards student-centred learning as class contact time is reduced, students and teachers alike are having to adjust to new methods of learning and teaching.

Essential Electronics is a series of textbooks intended for use by students on degree and diploma level courses in electrical and electronic engineering and related courses such as manufacturing, mechanical, civil and general engineering. Each text is complete in itself and is complementary to other books in the series.

A feature of these books is the acknowledgement of the new culture outlined above and of the fact that students entering higher education are now, through no fault of their own, less well equipped in mathematics and physics than students of ten or even five years ago. With numerous worked examples throughout, and further problems with answers at the end of each chapter, the texts are ideal for directed and independent learning.

The early books in the series cover topics normally found in the first and second year curricula and assume virtually no previous knowledge, with mathematics being kept to a minimum. Later ones are intended for study at final year level.

The authors are all highly qualified chartered engineers with wide experience in higher education and in industry.

R G Powell
Jan 1995
Nottingham Trent University

Contents

Preface

This textbook is suitable for electronic engineering students at about second year degree or final year HND level, who are studying the subject of power electronics for the first time. The material covered is sufficient for a 1 year course with average class contact of 3 hours per week, or 20 credit points of a modular course with 120 credit points per year.

Modern power electronics is the application of semiconductor devices to the control and conversion of electrical power. The availability of solid state power switches such as the thyristor and GTO, then BJT, Mosfet and IGBT power transistors, has created a very rapid expansion in power electronic applications from relatively low power control of domestic equipment to high power control of industrial processes and very high power flow control along transmission lines.

The table below indicates that power electronics development has been going on for some time, starting at the turn of the 20th century with the replacement of rotary converters by mercury arc rectifiers. This came about because of the need for efficient reliable direct current supplies for motor control and industrial processes, once it had been decided to use three-phase alternating current for transmission purposes. The arrival of the thyratron in the 1920s made possible the development of the first power electronic d.c. variable speed drive.

Historical landmarks

1880–1890	Electric light, DC generator and motor, alternator and synchronous motor
1890–1900	Induction motor, large transformers, rotary converters
1900–1920	Three-phase transmission, mercury arc rectifier, diode and triode
1920–1940	Thyratron, klystron and magnetron
1940–1960	Transistor, thyristor and triac
1960–	Mosfet, MCT and IGBT etc.

Power electronic switch units are now available in ratings from the general purpose, able to control, for example, 60 V at 10 A, through to modules controlling 250 kV at 1000 A. These switches are arranged in controller, converter and inverter circuits able to condition the power supply into the form required by the load. The switches can be connected in series to increase voltage handling capability, and in parallel to improve current handling.

The power converter, or power conditioner, is a connection of power swtiches into a topology which can rectify or invert, regulate and control the power flow through the system.

Power electronic systems are being installed throughout the world using switches of ever-increasing ratings. Some examples of large systems in the UK are:

1 The cross-channel link which connects together the a.c. electricity supply systems of England and France via a high voltage d.c. link to enable power to be transferred in either direction. The d.c. side is at $\pm$ 270 kV, 2000 MW and the a.c. side is 400 kV at 50 Hz. The switches in the inverter are thyristor units.

2 The BR Maglev system at Birmingham, which is a railway linking together the airport and the exhibition centre using trains with magnetic levitation and linear induction motors. The levitation magnets are supplied from 600 V d.c. via a 1 kHz chopper. Variable frequency for the induction motor is obtained from a PWM (pulse-width-modulated) transistor inverter with a 600 V d.c. link.

3 The Eurostar locomotive, on the channel tunnel, uses induction motors driven by GTO inverters with more than 1 MW of power available for driving the motors. In France and in the tunnel, the d.c. link voltage is 1900 V. In England the d.c. link voltage is 750 V.

The above applications are examples of spectacular power electronic developments. However, on a more everyday level, the power electronic industry is expanding and developing. In 1990, in one area alone, that of variable speed drives, about £80 million worth of business was done in the UK, roughly equally shared between a.c. and d.c. drives.

It seems a reasonable assumption to make that all electrical and electronic engineering students will, during their studies, and subsequently in their career, need to have background knowledge of power electronic theory. This book should meet that need as an introduction to the subject and as an indication of more advanced study areas.

D Fewson
Middlesex University
Jan 1998

Acknowledgements

This book is based on lectures given to students at Middlesex University over a period of about 10 years. Over this period I have been influenced by the many excellent textbooks referred to in the 'power electronic textbooks' section on page 185. In particular, I would like to thank C. Lander, M. Raschid and R. Ramshaw who, through their books, have been indirectly responsible for the success of a large number of Middlesex students. I must also thank the students themselves for checking the solutions to a number of problems used in this book, which have appeared on my tutorial sheets and examination papers. Thanks are also due to Ray Powell, the technical editor of the Arnold *Essential Electronics* series, for suggesting that I might like to contribute to the series.

As the title indicates, the book is an introduction to the subject of power electronics. The material is basic to the understanding of the subject and will form a platform from which a deeper study can be undertaken.

Mention has been made in the text of CAD circuit simulation packages to assist in the understanding of the operation of the power electronic circuits. The recommended size of the book has allowed only cursory inclusion of this area, but a course on power electronics at this level should include some circuit simulation. Two very good simulation packages are Microcap 5 produced by Spectrum, and ICAP/4 produced by 'intusoft'. Both manufacturers provide free introductory software ideal for student use.

Finally, I would like to thank Middlesex University for the use of their resources in preparing the material contained in this course book.

1 The power electronic system

1.1 INTRODUCTION

A power electronic system will consist of a power source, filtering, a power converter, a load and a control circuit. The block diagram is shown in Fig. 1.1.

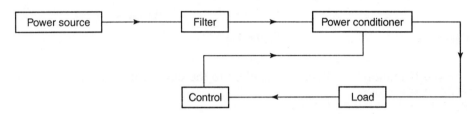

Figure 1.1

The *power source* could be three-phase, or single-phase, a.c. mains or it could be a portable supply such as a d.c. battery.

A *filter* may be necessary to prevent any harmonics generated by the converter from being fed back to the mains or from being radiated into space.

National and international standards for electromagnetic compatability (EMC) are now legally binding on manufacturers of power electronic equipment, e.g. EMC directive 89/336/EEC, IEC552–2, ECM regulations 1994, No. 3080 (HMSO).

The *control* circuit monitors the condition at the load, compares this with preset values and then adjusts the converter drive as necessary.

The *power conditioner* is an arrangement of semiconductor devices all operating in the switching mode. This means that the device is switched from cut-off to saturation ('off' to 'on') by the application of gate, or base, drive pulses. The ideal switch would have full voltage across it when 'off', and zero voltage across it when 'on'.

1.2 SWITCHING CHARACTERISTICS

The practical switch departs from the ideal in the manner shown in Figs 1.2 and 1.3. In this case a thyristor has been used as the switch, but a power transistor would have a similar switching characteristic.

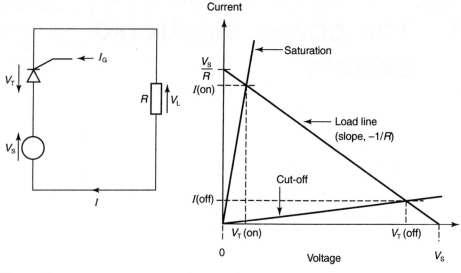

Figure 1.2   Figure 1.3

From Kirchhoff's voltage law applied to the circuit of Fig. 1.2

$$V_S = V_T + V_L = V_T + (I \times R)$$

where V_T is the voltage drop across the thyristor, V_L is the load voltage, I is the circuit current and R is the load resistance. Ideally when the switch is 'off', $V_T = V_S$ and $V_L = 0$, and when the switch is 'on', $V_T = 0$ and $V_L = V_S$. The switching characteristic in Fig. 1.3 shows how the practical switch differs from the ideal when the thyristor is switched on by the application of a gate pulse. The imperfections of the switch have been exaggerated for clarity.

1.3   POWER SWITCHES

Commonly available power switches are given in Table 1.1; this is not exhaustive and others are obtainable. Much research and development is going on and new devices will appear; also the power handling capability of existing devices is improving year by year. The forward voltage drop figures should be taken as a guide only, as this will depend on the gate or base drive values and on the rating.

1.4   CHOICE OF POWER SWITCH

Which of the power switches is chosen will be determined by cost, availability of rating to suit the requirement and the ease with which it can be turned 'on' and 'off'.

Table 1.1

Device	Turn-on	Turn-off	Rating (upper)	Forward voltage drop	Switching time (μs)
Thyristor (controlled rectifier)	Short duration gate pulse	Zero current or voltage reversal	1200 V/1500 A	1.5 V	20
Triac (bidirectional thyristor)	As thyristor	As thyristor	1200 V/300 A	1.7 V	20
GTO (controlled rectifier)	As thyristor	Reverse voltage gate pulse	1200 V/600 A	2.2 V	25
Power transistor (BJT)	Application of base current	Removal of base current	400 V/250 A	1.1 V	10
Power transistor (Darlington)	As BJT but lower base current	As BJT	900 V/200 A	1.5 V	40
Power transistor (Mosfet)	Application of gate voltage	Removal of gate voltage	600 V/40 A	1.2 V	>1
IGBT	Application of gate voltage	Removal of gate voltage	1200 V/50 A	3.0 V	>0.5

Mosfets and IGBTs have the simplest driving requirements; they are voltage controlled and the gate current is virtually zero during the 'on' period. However, they lack the reverse blocking capability which make the thyristor, Triac and GTO so suitable for a.c. mains power applications. With d.c. link inverters, the d.c. side means that turning off thyristors requires a forced commutation circuit, and GTOs are better.

However, if Mosfets are available with the correct rating then these, with reverse conducting diodes for inductive loads, would be a simpler choice.

The future of power electronics will almost certainly see the increasing use of 'application specific integrated circuits' (ASICs), at least for volume production. ASICs will combine switching and control requirements in a single module.

1.5   POWER CONDITIONER

Depending on the type of source and the type of load, the power conditioner, or converter, falls into the following categories:

- a.c.–d.c. controlled rectifiers

- d.c.–d.c. choppers

- a.c.–a.c. controllers

- d.c.–a.c. inverters

The a.c. to d.c. controlled rectifier provides a variable d.c. load voltage from a fixed voltage and frequency a.c. source. In the UK, the single-phase a.c. line to neutral voltage is nominally 240V, 50Hz. The three-phase a.c. line to line voltage is nominally 415V, 50Hz.

The d.c. to d.c. chopper provide variable d.c. load voltage from a fixed d.c. source voltage, typically a battery.

The a.c. to a.c. controllers provide variable a.c. load voltage from a fixed a.c. source voltage at constant frequency.

The d.c. to a.c. inverters produce a variable a.c. voltage and frequency from a fixed voltage d.c. source.

1.6   ANALYSIS OF POWER CONVERTER OPERATION

The level of mathematics required to solve some of the converters' Kirhhoff's law time-varying equations may not yet have been fully covered by students at the start of their second year. However, this need not prevent the understanding of the operation of the power electronic circuit, nor prevent solution of problems on the circuit's behaviour. Most of the solutions of the equations are worked through in the book and the important equation on which performance depends is enclosed in a box, e.g.

$$V_{av} = \sqrt{2}V_S(1 + \cos \alpha)/\pi$$

Worked, self-test and tutorial examples will examine circuit operation using these equations.

There are software circuit simulation packages available which enable circuits to be simulated and analysed using a computer. These give both numerical and graphical solutions of the circuit behaviour. They can save an enormous amount of time in predicting performance and can prevent costly mistakes in prototype circuits. Some of the solutions in this book are confirmed using simulation methods.

Examples of simulation software are PSPICE, MICROCAP, Electronics Workbench and SABRE. There are other simulation packages available. It is fair to say that you will need to become proficient in both formal methods and simulation if you are to achieve a good level of competence in the power electronics field. The full industrial version of the simulation software can be quite expensive, but some cheaper student editions are available, and also some free evaluation software can be obtained.

1.7 APPLICATIONS OF POWER ELECTRONICS

It is astonishing to realise that there is hardly a home, office block, factory, car, sports hall, hospital or theatre without an application, and sometimes many applications, of power electronic equipment.

Some typical applications are given below:

- industrial processes in the chemical, paper and steel industries;

- domestic and theatre lighting;

- motor drives from food mixers and washing machines through to lifts and locomotives such as 'Eurostar';

- power supplies for laboratories and uninterruptible power for vital loads;

- generation and transmission control;

- heating and ventilating of homes and office blocks.

2 DC to DC choppers

2.1 STEP-DOWN CHOPPERS

The power source for the chopper could be a battery or a rectified a.c. The purpose of the chopper is to provide variable d.c. voltage from a fixed voltage d.c. source. Applications of choppers are in drives for electric vehicles, in the d.c. link for variable frequency inverters and in switched mode power supplies.

The power switch can be a power transistor such as a BJT, Darlington, Mosfet or IGBT. These require base, or gate driver, circuits to turn the switch on, and they turn off when the driver pulse is removed. The power switch could also be a thyristor (SCR), but on d.c. a separate 'turn off' or forced commutation circuit is required and this complication tends to rule out the thyristor for all but very high power circuits.

2.2 CHOPPERS WITH RESISTIVE LOADS

Typical chopper circuits are shown in Figs 2.1–2.4 with a load resistance R. The switch is assumed to be ideal with zero voltage across it when 'on', and full battery voltage across it when 'off'. Care must be taken if the power source and the base, or gate driver, circuit have a common terminal. If this is the case, Figs 2.2–2.4 would be fine, but Fig. 2.1 would need base driver isolation with an opto-isolator.

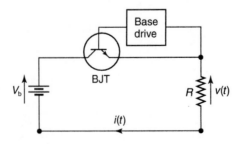

Figure 2.1

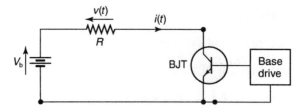

Figure 2.2

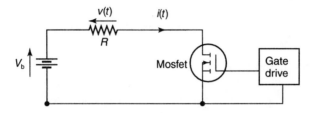

Figure 2.3

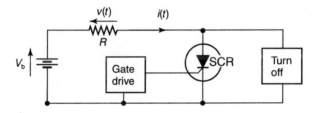

Figure 2.4

In the pulse-width-modulated (PWM) chopper, the frequency is chosen to be some convenient value, and the ratio of time-on (t_n) to time-off (t_f) is adjusted to produce the required value of load voltage. Whatever the type of power switch used, the load voltage waveform will be the same, as shown in Fig. 2.5.

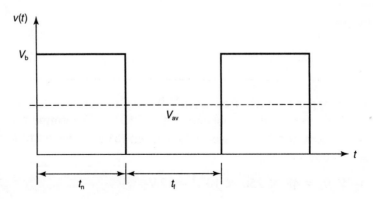

Figure 2.5

Average load voltage

From Fig. 2.5, the average load voltage

V_{av} = area under curve/length of base

$$= (V_b t_n)/(t_n + t_f)$$

Alternatively, by integration

$$V_{av} = \frac{1}{(t_n + t_f)} \int_0^{t_n} V_b \partial t \; t$$

$$= (V_b t_n)/(t_n + t_f)$$

The periodic time of the voltage

$$T = (t_n + t_f)$$

Hence

$$V_{av} = (V_b t_n)/T = V_b t_n f \qquad (2.1)$$

Also

$$I_{av} = V_{av}/R \qquad (2.2)$$

Root-mean-square load voltage (V_{rms})

$$V_{rms} = \sqrt{\frac{1}{T} \int_0^{t_n} V_b^2 \delta t} = V_b \sqrt{\frac{t_n}{T}} = \boxed{V_b \sqrt{t_n f}} \qquad (2.3)$$

For resistive load, R

$$I_{rms} = V_{rms}/R \qquad (2.4)$$

Example 2.1

A d.c. to d.c. chopper operates from a 48 V battery source into a resistive load of 24 Ω. The frequency of the chopper is set to 250 Hz. Determine the average and rms load current and load power values when (a) chopper on-time is 1 ms, (b) chopper on-time is 3 ms.

(a) $V_{av} = V_b f t_n = 48 \times 250 \times 10^{-3} = 12 V$

$I_{av} = V_{av}/R = 12/24 = 0.5 A$

$V_{rms} = V_b \sqrt{t_n f} = 48 \sqrt{0.25} = 24 V$

$$I_{\text{rms}} = V_{\text{rms}}/R = .24/24 = 1\,\text{A}$$

$$P = I_{\text{rms}}^2 \times R = 1 \times 24 = 24\,\text{W}$$

(b) $V_{\text{av}} = V_{\text{b}}ft_{\text{n}} = 48 \times 250 \times 3 \times 10^{-3} = 36\,\text{V}$

$\quad I_{\text{av}} = V_{\text{av}}/R = 36/24 = 1.5\,\text{A}$

$\quad V_{\text{rms}} = V_{\text{b}}\sqrt{t_{\text{n}}f} = 48\sqrt{0.75} = 41.6\,\text{V}$

$\quad I_{\text{rms}} = V_{\text{rms}}/R = 41.6/24 = 1.73\,\text{A}$

$\quad P = I_{\text{rms}}^2 \times R = 1.73^2 \times 24 = 71.8\,\text{W}$

2.3 CHOPPERS WITH INDUCTIVE LOADS

When the load contains inductance as well as resistance then, when switching occurs, circuit currents cannot change their values instantaneously due to energy storage in the inductor. At switch-on, the current grows exponentially as energy is stored in the inductor, while at switch-off the energy is dissipated and the power electronic switch must be protected against the possibility of high inductive voltage rise causing damage to the switch. This is normally done by connecting a diode in parallel across the load. The diode is called a free-wheeling diode. At switch-off, the inductive voltage will forward bias the diode, allowing exponential decay of the current around the loop of load and diode. A typical circuit is shown in Fig. 2.6.

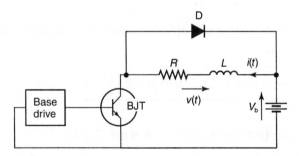

Figure 2.6

Analysis of circuits with inductive loads is often carried out using the Laplace transformation. This changes the normal time-domain circuit into a frequency-domain circuit using the Laplace operator s. The resulting equation in s can be manipulated algebraically into a form that can be returned to the time domain using a table of transform pairs. The full Laplace transformation solutions are given below. If you have not yet covered the theory, don't worry since we will be using final equations to determine performance of the chopper and for the solutions to problems.

In the following analysis, the switch is considered ideal. Also the assumption is that the current is continuous with a load current of I_0 at switch-on, and a current of I_1 at switch-off.

Chopper on-period (see Figs 2.7 and 2.8)

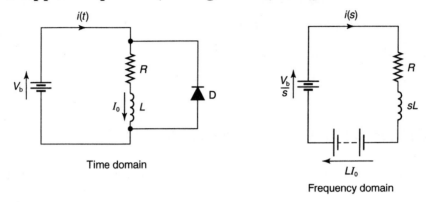

Figure 2.7   Figure 2.8

From the circuit in Fig. 2.8

$i(s) = ((V_b/s) + LI_0)/(R + sL)$

$\quad = V_b/s(R + sL) + LI_0/(R + sL)$

This simplifies to

$i(s) = V_b a/Rs(s + a) + I_0/(s + a)$

where $a = (R/L)$. From a table of Laplace transform pairs, this transforms back to the time domain as

$$i(t) = (V_b/R)(1 - \exp(-Rt/L)) + I_0\exp(-Rt/L)$$ (2.5)

Chopper off-period (see Figs 2.9 and 2.10)

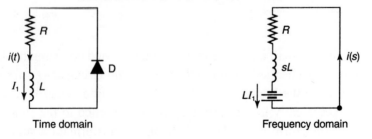

Figure 2.9   Figure 2.10

$i(s) = LI_1/(R + sL) = I_1/(s + a)$

where $a = R/L$. Hence

$$i(t) = I_1 \exp(-Rt/L)$$ (2.6)

Chopper waveforms (see Fig. 2.11)

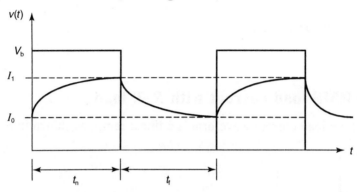

Figure 2.11

Example 2.2 _____

A d.c. to d.c. chopper has an inductive load of 1Ω resistance and 10mH inductance. Source voltage is 24V. The frequency of the chopper is set to 100Hz and the on-time to 5ms. Determine the average, maximum and minimum load currents.

Solution

$V_{av} = V_b f t_n = 24 \times 100 \times 5 \times 10^{-3} = 12\text{V}$

$I_{av} = V_{av}/R = 12/1 = 12\text{A}$

During the on-period

$i(t) = (V_b/R)(1 - \exp(-Rt/L)) + I_0\exp(-Rt/L)$

$Rt/L = 1 \times 5 \times 10^{-3}/10 \times 10^{-3} = 0.5$

Now at the end of the on-period, $i(t) = I_1$:

$I_1 = 24(1 - \exp(-0.5)) + I_0\exp(-0.5) = 9.44 + 0.607\,I_0$ (a)

During the off-period

$i(t) = I_1\exp(-Rt/L)$

At the end of the off-period, $Rt/L = 1 \times 5 \times 10^{-3}/10 \times 10^{-3} = 0.5$ and $i(t) = I_0$; hence

$$I_0 = I_1\exp^{-0.5} = 0.607\,I_1 \tag{b}$$

Substitute (b) into (a):

$I_1 = 9.44 + 0.607(0.607)\,I_1$

$\therefore I_1 = 14.95\,\text{A}$

$I_0 = 0.607\,I_1 = 0.607 \times 14.95$

$\therefore I_0 = 9.07\,\text{A}$

Chopper RMS load current with *R–L* load

Assume that the load current waveform is a linear ramp, as shown in Fig. 2.12.

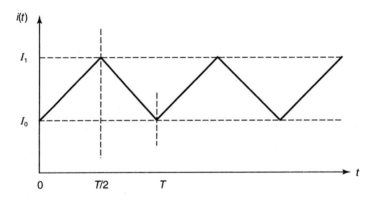

Figure 2.12

Now

$i(t) = I_0 + kt$

where $k = (I_1 - I_0)/(T/2)$. Hence

$i^2(t) = I_0^2 + 2I_0kt + k^2t^2$

$I_{\text{rms}} = \sqrt{(2/T) \int_0^{T/2} i^2(t)\partial t}$

$\qquad = \sqrt{(2/T)\,[I_0^2t + 2I_0kt^2/2 + k^2t^3/3]_0^{T/2}}$

$\qquad = \sqrt{I_0^2 + kT/2 + k^2T^2/12}$

$\qquad = \sqrt{I_0^2 + I_0(I_1 - I_0) + (I_1 - I_0)^2/3}$

$$\boxed{I_{\text{rms}} = \sqrt{I_0I_1 + (I_1 - I_0)^2/3}} \tag{2.7}$$

Equation (2.7) gives the rms value of the load current in Example 2.2 as

$$I_{rms} = \sqrt{(14.95 \times 9.07) + (14.95 - 9.07)^2/3} = 12.13\,\text{A}$$

Circuit simulation

A PSPICE circuit simulation has been run for the chopper circuit in Example 2.2 using free evaluation software provided by the MicroSim Corporation. The results are given in Fig. 2.13, and they show $T_0 = 9.06\,\text{A}$, $T_1 = 14.95\,\text{A}$ in close agreement with the calculations.

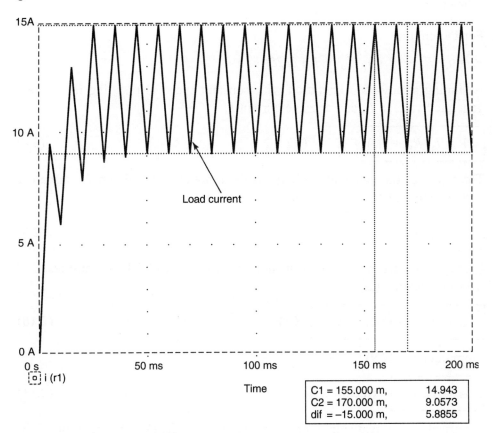

C1 = 155.000 m,	14.943
C2 = 170.000 m,	9.0573
dif = −15.000 m,	5.8855

Figure 2.13 Chopper with *L-R* load

2.4 DC SERIES MOTOR

A d.c. motor consists of a stationary field winding (R_f, L_f), to produce the pole flux, and a rotating armature winding (R_a, L_a) through which the supply current flows to produce torque and rotation.

The circuit arrangement of d.c. series motor is given in Fig. 2.14.

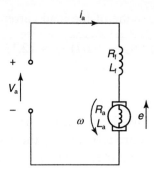

Figure 2.14

For the circuit

$$V_a = i_a (R_a + R_f) + (L_a + L_f) \, \partial i_a/\partial t + e$$

Assuming steady-state operation with constant speed ω resulting in constant generated voltage E, and average current I_a

$$V = I_a (R_a + R_f) + E \qquad \text{(V)} \tag{2.8}$$

The value of the generated voltage in the armature depends on the armature voltage constant k_v, the speed and current:

$$E = k_v \, \omega \, I_a \qquad \text{(V)} \tag{2.9}$$

k_v may be quoted by the motor manufacturer or determined by motor tests.
  From equation (2.8)

$$I_a = (V_a - E)/(R_a + R_f) \qquad \text{(A)} \tag{2.10}$$

From equation (2.9)

$$\omega = E/k_v I_a \qquad \text{(rad/s)} \tag{2.11}$$

Electrical power available to create torque is $P_e = EI_a$ (W); hence torque T is

$$T = EI_a/\omega = k_v I_a^2 \qquad \text{(Nm)} \tag{2.12}$$

2.5   SERIES MOTOR CHOPPER DRIVE

A basic chopper drive system using a BJT as the switching element is shown in Fig. 2.15 where R and L are the combined armature and field resistances and inductances, respectively.

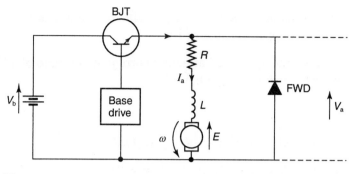

Figure 2.15

Assuming that during the brief 'on' and 'off' periods of the BJT both motor speed and generated voltage remain constant, the voltage and current waveforms will be shown in Fig. 2.16.

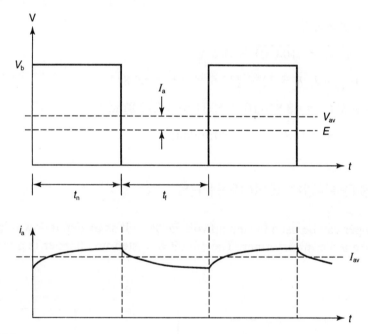

Figure 2.16

The drive performance can be evaluated using equations (2.8)–(2.12). Remember we are working in average values and the average armature voltage, V_{av}, is related to the steady battery voltage, V_b, by the fundamental chopper equation

$$V_{av} = V_b t_n f \qquad (2.13)$$

Example 2.3

A d.c. series motor chopper drive has the following parameters: battery voltage = 96 V, $(R_a + R_f) = 0.1\,\Omega$, $k_v = 10\,\text{mV/A-rad/s}$, chopper frequency = 125 Hz.

(a) Calculate the armature speed and torque with an average armature current of 100 A and a chopper on-time of 6 ms.

(b) Determine the motor speed and armature current at a torque of 10 Nm with on-time and off-time as in (a) above.

Solution

(a) $V_{av} = V_b t_n f = 96 \times 6 \times 10^{-3} \times 125 = 72\,\text{V}$

  $E = V_a - I_{av}R = 72 - (100 \times 0.1) = 62\,\text{V}$

  $\omega = E/k_v I_{av} = 62/(10 \times 10^{-3} \times 100) = 62\,\text{rad/s}$

  $N = \omega \times (60/2\pi) = 592\,\text{rev/min}$

  $T = k_v I_{av}^2 = 10^{-2} \times 100^2 = 100\,\text{Nm}$

(b) Since $T = k_v I_{av}^2$

  $I_{av} = \sqrt{T/k_v} = \sqrt{10/0.01} = 31.6\,\text{A}$

  $E = V_{av} - I_{av}R = 72 - (31.6 \times 0.1) = 68.84\,\text{V}$

  $\omega = E/k_v I_{av} = 68.84/(10^{-2} \times 31.6) = 217.8\,\text{rad/s}$

  $N = 217.8 \times (60/2\pi) = 2080\,\text{rev/min}$

2.6 STEP-UP CHOPPERS

The chopper can be used to step up voltage as well as to step it down; it behaves in fact like a d.c. transformer. The circuit of a step-up chopper is given in Fig. 2.17.

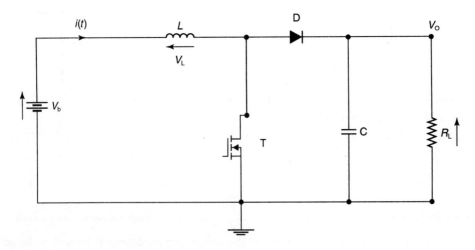

Figure 2.17

When the transistor is switched on, current ramps up in the inductor and energy is stored. When the transistor is switched off, the inductor voltage reverses and acts together with the battery voltage to forward bias the diode, transferring energy to the capacitor. When the transistor is switched on again, load current is maintained by the capacitor, energy is stored in the inductor and the cycle can start again. The value of the load voltage is increased by increasing the duty cycle or the on-time of the transistor.

During the transistor on-period, assuming an ideal inductor and transistor

$V_b = V_L = L di/dt$

$V_b/s = (sL) i(s)$

$i(s) = V_b/s^2 L$

$i(t) = (V_b/L)t = kt$

Over an on-time of t_n, the change of battery current will be

$\Delta i = kt_n$

During the off-period, t_f

$V_0 = V_b + V_L = V_b + L(di/dt) = V_b + L(\Delta i/\Delta t)$

$\quad = V_b + L(V_b/L)t_n/t_f = V_b(1 + t_n/t_f)$

Let the duty cycle $D = t_n T$, and $t_f = T - t_n = T(1 - D)$. Now

$V_0 = V_b(1 - DT/T(1 - D))$

which simplifies to

$$V_0 = V_b/(1 - D) \tag{2.14}$$

In an ideal circuit, $V_b I_b = V_0 I_0$. Therefore, from equation (2.14)

$I_b = (V_0/V_b)I_0 = I_0/(1 - D)$

$I_b = I_b + \Delta i/2$

$I_{b0} = I_b - \Delta i/2$

When the steady variation of battery current has been reached, the variation will be between I_0 and I_1, as shown in Fig. 2.18.

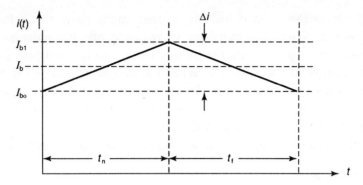

Figure 2.18

Example 2.4

A step-up chopper similar to that shown in Fig. 2.17 is to deliver 3A into the 10Ω load. The battery voltage is 12V, $L = 20\,\mu H$, C = 100 μH and the chopper frequency is 50kHz. Determine the on-time of the chopper, the battery current variation and the average battery current.

Solution

$V_0 = I_0 R_L = 3 \times 10 = 30\,V$ ·

From equation (2.14), $V_b = V_0(1 - D)$. Hence

$1 - D = V_b/V_0 = 12/30 = 0.4$

Hence $D = 0.6$.

Periodic time $T = 1/f = 1/50 \times 10^3 = 20\,\mu s$

Time-on $t_n = DT = 0.6 \times 20 \times 10^{-6} = 12\,\mu s$

$k = V_b/L = 12/20 \times 10^{-6} = 0.6 \times 10^6$ A/s

$\Delta i = kt_n = 0.6 \times 10^6 \times 12 \times 10^{-6} = 7.2\,A$

Average battery current

$I_b = I_0/(1 - D) = 3/(1 - 0.6) = 7.5\,A$

$I_{b1} = I_b + \Delta i/2 = 7.5 + 7.2/2 = 11.1\,A$

$I_{b0} = I_b - \Delta i/2 = 7.5 - 3.6 = 3.9\,A$

A PSPICE simulation of the above circuit is given in Fig 2.19, where it can be seen that the approximate calculations are not too far out.

In the simulation, after transients have died away, $V_0 = 27\,V$ compared to the predicted 30V. $I_{b0} = 3.8\,A$ compared to the calculated value of 3.9A. I_{b1} is about 10.4A compared to the calculated value of 11.1A.

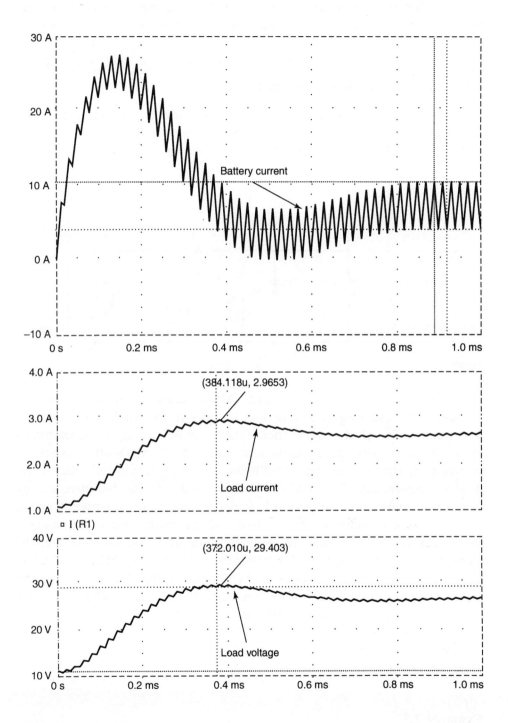

Figure 2.19

Remember that the calculations assumed ideal components, and the simulation will use library components with parameters close to those encountered in practice.

DC motor drive with regeneration

If a step-up chopper is included on a series motor variable speed drive, the drive can be made to regenerate back to the battery. This arrangement would allow motor braking. The circuit required for this is shown in Fig 2.20.

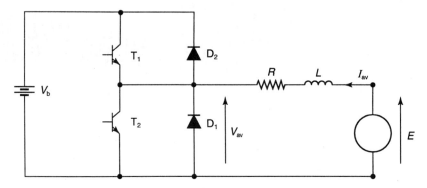

Figure 2.20

T_1 and D_1 comprise the step-down chopper allowing normal control of forward speed and torque. The step-up chopper is formed from T_2 and D_2. This would be switched into operation if the motor was acting as a generator, as in the case of a vehicle freewheeling down hill. With T_1 and D_1 off, T_2 would be switched on, armature current would build up to a predetermined value in the direction shown, and T_2 would then be switched off. The inductive energy stored in L would cause the voltage V_a to rise to keep armature current flowing, D_2 would then becomes forward biased and the battery would be charged. Remember the current is ramping up and down in the T_2 'on' and 'off' periods. Current would then decay to a predetermined limit and T_2 would be switched on again. Control of braking would be by control of the duty cycle. On the assumption of constant average armature current, the equation already derived for the step-up chopper would apply, i.e.

$$V_a = V_b(1 - D)$$

Example 2.5

A small electric vehicle is powered by a 96 V battery and uses a series wound motor. It has a step-down chopper for forward drive and a step-up chopper for regenerative braking. The total armature resistance is $0.1\,\Omega$, and the armature voltage constant is 10m V/A-rad/s. Descending a hill in regeneration mode the motor speed is 1000 rev/min and the armature current is a constant average value of 80 A.

Calculate the value of the duty cycle required and the available braking power.

Solution

$$E = \omega k_v I_{av} = (1000 \times 2\pi/60) \times 10^{-2} \times 80 = 84\,V$$

$$V_{av} = E - I_{av}R = 84 - (40 \times 0.1) = 80\,V$$

$$V_{av} = V_b\,(1 - D)$$

$$D = 1 - (V_{av}/V_b) = 1 - (80/96) = 0.17$$

$$P = V_{av} \times I_{av} = 80 \times 80 = 6.4\,kW$$

2.7 TURNING ON POWER SWITCHES

If the power switch is a BJT, the current pulse amplitude on the base depends on the collector load current. Current amplification factors for high power BJTs are not that large and could be as low as 10. In this case a buffer transistor between the driver timing circuit and the power transistor would be required. One solution is to use a power Darlington transistor perhaps with a current gain of 200. In this case a load of 40 A would require a base current of 200 mA to switch on, within the reach of some integrated circuit (IC) modules.

If the power switch is a Mosfet then gate current requirements are almost negligible, of the order of nA, and can be driven directly by CMOS logic.

One useful IC module for timing and base drive is the 555 timer. It can be connected as a pulse-width modulator (PWM) and can switch 200 mA on its output.

The external components required to program the timer are shown in Fig. 2.21.

Let

R_1 and the top half of $R_2 = R_a$

R_3 and the bottom half of $R_2 = R_b$

The frequency of the timer is given by

$$f(Hz) = 1.44/(R_a + R_b)C_1$$

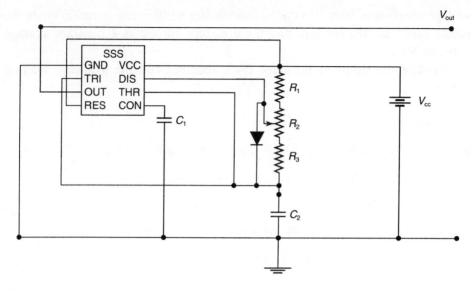

Figure 2.21

An Electronics Workbench simulation of a timer acting as a base driver for a BJT in a chopper circuit is given in Fig. 2.22 for a 0.5 and a 0.2 duty cycle. The frequency is about 14 kHz. The top oscilloscope trace is V_{ce} of the BJT and the bottom is V_{out} of the timer.

2.8 TURNING OFF THYRISTOR CIRCUITS

To turn on a thyristor, a low voltage, short duration pulse is applied to the gate (typically 4 V, 100 μs). Once the thyristor is 'on', the gate loses control and the thyristor will only turn off when the load current falls virtually to zero, or the thyristor is reverse biased. The thyristor will turn off naturally with a.c. supplies as the voltage reverses, but no such reversal occurs with d.c. supplies and it is necessary to force a voltage reversal if turn-off is to occur. This process is called 'forced commutation'. One method of achieving forced commutation is parallel capacitor turn-off. A typical circuit is shown in Fig. 2.23.

Consider T1 to be 'on' and supplying load current to R_L; C has charged up with the bottom plate positive via R and T_1. When T_1 is to be turned off, T_2 is switched on, connecting the charged capacitor in reverse across T_1, and it prepares to turn off. Meanwhile C discharges and charges up in reverse through the supply, R_L and T_2. When T_1 is switched on again, T_2 is reverse biased and switches off. The cycle can now start over again. Consider T_2 switched on and T_1 switching off. The equivalent circuit is shown in the s domain in Fig 2.24.

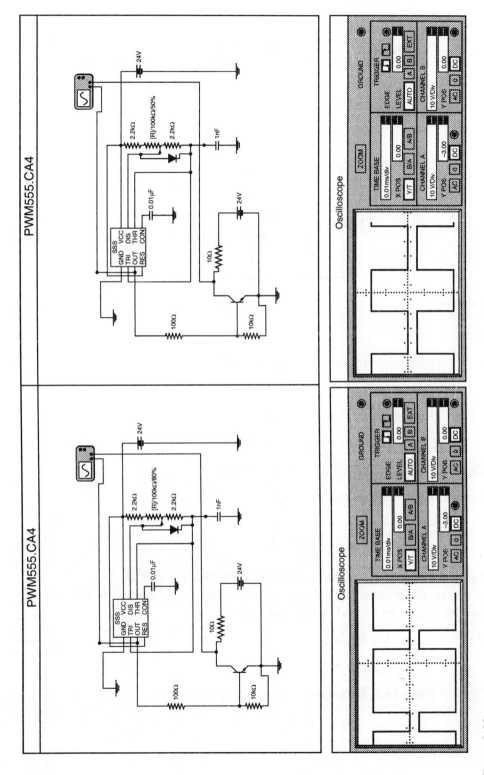

Figure 2.22

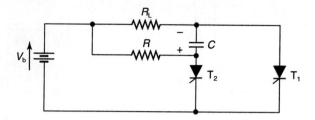

Figure 2.23

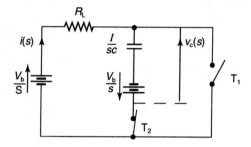

Figure 2.24

Now

$$i(s) = 2V_b/s(R_L + 1/sC) = 2V_b/R_L(s + 1/CR_L)$$

$$v_c(s) = (i(s) \times 1/sC) - V_b/s$$

$$= 2V_b/sCR_L(s + 1/CR_L) - V_b/s$$

$$= (2V_b/CR)(1/s(s + 1/CR) - V_b/s$$

This transforms back to the time domain as

$$v_c(t) = 2V_b(1 - \exp(-t/CR)) - V_b$$

$$= \boxed{V_b(1 - 2\exp(-t/CR))} \tag{2.15}$$

The waveform of this capacitor voltage is shown in Fig 2.25.

The thyristor must have been turned off by t_0, otherwise it will become forward biased again and not turn off. Thus

$$V_b(1 - 2\exp(-t_0/CR)) = 0$$

$$\exp(-t_0/CR) = 1/2$$

$$\exp(+t_0/CR) = 2$$

$$t_0 = CR \log 2$$

$$= 0.7 \ CR$$

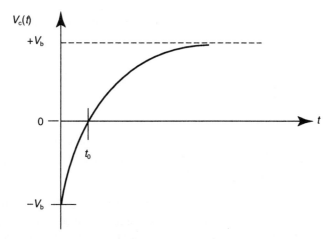

Figure 2.25

The size of capacitor required for communication is therefore

$$C = t_0/0.7R = 1.43\ t_0 I/V_b \qquad (2.16)$$

where t_0 is the thyristor turn-off time and I is the full-load current.

Example 2.5_____

What value of capacitor is required to force commutate a thyristor with a turn-off time of $20\,\mu s$ with a 96 V battery and a full-load current of 100 A?

Solution

$C = 1.43 t_0 I/V$

$\quad = 1.43 \times 20 \times 10^{-6} \times 100/96 = 30\,\mu F$

2.9 SELF-ASSESSMENT TEST

1 Sketch the circuit diagram of a Mosfet d.c. to d.c. chopper supplying variable voltage to a resistive load. With the aid of a voltage waveform diagram, obtain an expression for the average load voltage.

2 Draw voltage and current waveforms for a d.c. to d.c. chopper driving an R–L load. Assume continuous load current flow between switching periods.

3 Draw a circuit diagram of a d.c. series motor chopper drive, and by means of voltage and current waveforms show the behaviour of the circuit.

4 Write down the steady-state performance equations of a d.c. series motor

chopper drive for armature and generated voltage, armature current, speed and torque.

2.10 PROBLEMS

1 A 60V chopper drives a resistive load of 12Ω. The chopper frequency is set to 1kHz. Calculate the values of average and rms load currents and load power for a turn-on time of (a) 0.2ms, (b) 0.6ms.

2 A transistor PWM chopper supplies power to a parallel combination of inductive load and freewheel diode. The battery supply is 24V, the load has a resistance of 1Ω and inductance of 10mH, and the chopper frequency is 100Hz. Determine average load voltage, and minimum and maximum load currents for a turn-on time of (a) 8ms, (b) 2ms.

3 A 5kW d.c. series motor is to be used in a variable-speed drive controlled by a chopper. The motor field and armature resistances both equal 0.2Ω. With light load on the motor shaft, the armature runs at 3000rev/min and takes 8A from a 48V supply. Determine the armature voltage constant k_v.

   With the supply above the frequency above the chopper set to 250 Hz, and the on-time to 3.5ms, the average armature current is found to be 80A. Calculate the values of armature speed and torque.

4 A d.c. series motor is controlled by a chopper. Drive parameters are as follows:

   $(R_a + R_f) = 0.1\Omega$

   $k_v = 10mV/A\text{-rad/s}$

   $V_b = 96V$

   $f = 125Hz$

   (a) With an armature torque of 5Nm, the speed is 1500rev/min. What are the values of armature current and turn-on time?

   (b) The shaft torque is increased to 20Nm and the chopper on-time increased to hold the armature speed at 1500rev/min. What is the new armature current and on-time?

5 A step-up chopper is to provide a 48V across a 12Ω load from a 12V d.c. supply. The chopper inductance is 10μH and the chopper frequency is 100kHz. Determine the duty cycle required and the variation of battery current during switching, and the maximum and minimum battery currents.

3 AC to DC thyristor converters

3.1 INTRODUCTION

In the UK, generation and transmission of electrical power are by means of alternating current. The power stations use synchronous generators, called alternators, to generate at about 11kV, or higher, at a frequency of 50Hz. The voltage is then stepped up using auto-transformers to a value considered economic for transmission; this can be 132kV (the grid), or 275 and 400kV (the super grid).

Grid switching stations are used to interconnect the various voltage levels. Grid supply points have transformers to reduce the voltage to 33kV. Sub-stations reduce the voltage to 415/240V for distribution to homes and factories (there are also other voltage levels).

Some electrical equipment can use a.c. directly, e.g. lamps, space and water heating, cookers, fans, drills, vacuum cleaners etc. Other applications require that a.c is changed to d.c. These include radio and TV sets, computers, battery chargers, TTL and CMOS logic circuits, laboratory power supplies, public transport traction drives, high voltage d.c. links, etc.

Many different d.c. levels are required depending on the applications, e.g.:

- 5V for TTL and CMOS logic

- $\pm$ 15V for operational amplifiers

- 5, 12, 15, 24, 30 and 60V for laboratory power supplies

- 12, 24, 48 and in excess of 100V for battery charging

- 220V for d.c. motors

- 750V for underground trains

- 1900V for the Channel Tunnel

- 25kV for the EHT on cathode ray tubes

- $\pm$250kV for the high voltage d.c. link.

The process of changing a.c. into d.c. is called *rectification*. Where the application requires fixed voltage d.c., the switching element is a diode. Where the application requires variable voltage d.c., controlled rectifiers are used.

3.2   SINGLE-PHASE HALF-WAVE CONTROLLED RECTIFIER

The simplest controlled rectifier uses a single device, such as a thyristor, to produce variable voltage d.c. from fixed voltage a.c. mains. The circuit arrangement is shown in Fig. 3.1.

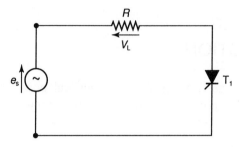

Figure 3.1

In this arrangement

$$e_s = E_m \sin \omega t = E_m \sin \theta = \sqrt{2}\, E_s \sin \theta$$

where E_m and E_s are the maximum and rms value of the supply voltage, respectively.

The thyristor is turned on in the positive half-cycle, some time after supply voltage zero, by the application of a gate pulse with delay angle α. In the negative half-cycle, the thyristor is reverse biased and cannot switch on. The larger the delay angle, the smaller is the average load voltage. Voltage waveforms for two delay angles are shown in Fig. 3.2.

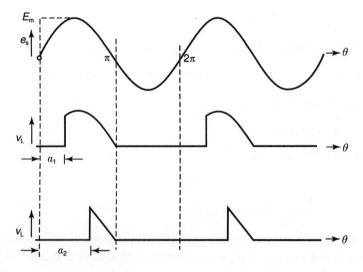

Figure 3.2

Average load voltage (V_{av})

Average load voltage is found by calculating the area under the voltage curve then dividing by the length of the base. For any delay angle a, the average load voltage is given by

$$V_{av} = (1/2\pi) \int_{a}^{\pi} E_m \sin \theta \, \delta\theta = (E_m/2\pi) \, [-\cos\theta]$$

Hence

$$\boxed{V_{av} = (E_m/2\pi) \, (1 + \cos a)} \tag{3.1}$$

$$\boxed{I_{av} = V_{av}/R} \tag{3.2}$$

Example 3.1 _____

A thyristor half-wave controlled converter has a supply voltage of 240 V at 50 Hz and a load resistance of 100 Ω. What are the average values of load voltage and current when the firing delay angle is (a) 30°, (b) 140°?

Solution

(a) From equation (3.1)

$$V_{av} = (E_m/2\pi) \, (1 + \cos a)$$
$$= (\sqrt{2} \times 240/2\pi) \, (1 + \cos 30°)$$
$$= 100.8 \, V$$

From equation (3.2)

$$I_{av} = V_{av}/R = 100.8/100$$
$$= 1.01 \, A$$

(b) $V_{av} = (\sqrt{2} \times 240/2\pi) \, (1 + \cos 140°)$
$$= 12.6 \, V$$
$$I_{av} = V_{av}/R = 12.6/100$$
$$= 126 \, mA$$

The Electronics Workbench simulation in Fig. 3.3 shows ammeter readings in close agreement. The top trace is the supply voltage, e_s, and the bottom trace is the load voltage, v_L. The behaviour of the thyristor gate circuit, consisting of the 10 to 120 kΩ resistor, the 100 nF capacitor and the ECG6412 Diac, to delay the thyristor switch-on will be described in Section 3.3.

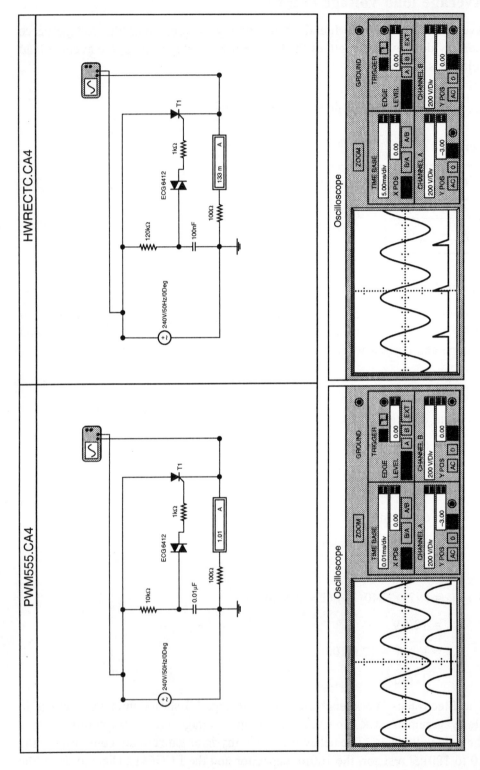

Figure 3.3

Root-mean-square load voltage, (V_{rms})

The square root of the average value of the square of the time-varying voltage gives the rms value:

$$V_{rms} = \sqrt{(1/2\pi) \int_{\alpha}^{\pi} (E_m \sin \theta)^2 \, \delta\theta}$$

Using the identity that $\sin^2 \theta = 0.5(1 - \cos2\theta)$

$$V_{rms} = E_m \sqrt{(1/2\pi) \int_{\alpha}^{\pi} 0.5(1 - \cos 2\theta) \, \delta\theta}$$

$$= (E_m/2) \sqrt{(1/\pi) \left[\theta - (\sin 2\theta)/2 \right]_{\alpha}^{\pi}}$$

$$= (E_m/2) \sqrt{(1/\pi) (\pi - \alpha + (\sin 2\alpha)/2)}$$

$$V_{rms} = (E_m/2) \sqrt{(1 - \alpha/\pi + (\sin 2\alpha)/2\pi)} \tag{3.3}$$

$$I_{rms} = V_{rms}/R \tag{3.4}$$

Example 3.2 _____

For the circuit in example 3.1, find the values of rms current, load power and overall power factor when (a) $\alpha = 30°$, (b) $\alpha = 120°$.

Solution

(a) From equation (3.3)

$$V_{rms} = (\sqrt{2} \times 240/2) \sqrt{(1 - 30/180 + (\sin 60)/2\pi)}$$

$$= 167.3 \, V$$

From equation (3.4)

$$I_{rms} = 167.3/100 = 1.673 \, A$$

$$P = I_{rms}^2 R = (1.673)^2 \times 100 = 280 \, W$$

Power factor $= P/V_s I_{rms} = 280/240 \times 1.673 = 0.697$

(b) $V_{rms} = (\sqrt{2} \times 240/2) \sqrt{(1 - 120/180 + (\sin 240)/2\pi)}$

$$= 75.04 \, V$$

$$I_{rms} = 75.04/100 = 0.75 \, A$$

$$P = I_{rms}^2 R = (0.75)^2 \times 100 = 56.3 \, W$$

Power factor $= P/V_s I_{rms} = 56.3/240 \times 0.75 = 0.31$

3.3   THYRISTOR TURN-ON

A simple gating, or firing circuit, consists of a *C–R* circuit to control the time to build up a particular voltage across the capacitor, and a Diac which will break-over at a voltage, typically 20–60 V, depending on the type. When the break-over voltage is reached, voltage drop across the Diac will fall to about 1.5 V. In the circuit in Fig. 3.4, the thyristor will turn on at about Diac break-over.

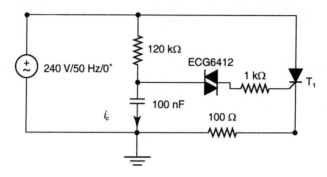

Figure 3.4

The Diac ECG6412 has a break-down, or switching voltage, of 63 V.

It can be shown that the current i_c in the *R–C* circuit is given by the expression

$$i_c = (E_m/|Z|) \sin(\omega t + \phi)$$

where

$$|Z| = \sqrt{R^2 + X_c^2}$$

and

$$\phi = \tan^{-1}(X_c/R)$$

The capacitor voltage

$$v_c = i_c X_c$$

lagging the current by 90°.

$$X_c = 1/\omega C$$

$$v_c = (E_m X_c/|Z|) \sin(\omega t + \phi - 90°)$$

Example 3.3 _____

In the circuit in Fig. 3.4, $C = 100\,\text{nF}$ and *R* is variable from 10 to 120 kΩ. Determine the range of firing angle delay available.

Solution

$X_c = 1/\omega C = 10^5/\pi = 31.83 \text{k}\Omega$

$V_m = \sqrt{2} \times 240 = 340 \text{V}$

(a) $R = 10 \text{k}\Omega$.

$|Z| = \sqrt{R^2 + X_c^2} = 10^4 \sqrt{(1 + 3.183^2)} = 33364 \Omega$

and

$\phi = \tan^{-1} (X_c/R) = \tan^{-1} (3.183) = 72.6°$

$i_c = (E_m/|Z|) \sin(\omega t + \phi)$

$\quad = (340/33364) \sin (a + 72.6°)$

The capacitor voltage

$v_c = (E_m X_c/|Z|) \sin (\omega t + \phi - 90°)$

$\quad = 324.36 \sin (a - 17.4°) = 63$

where 63 V is the Diac break-over voltage. Hence

$\sin(a - 17.4°) = 0.1942$

$(a - 17.4°) = \sin^{-1} (0.1942) = 11.2°$

$a = 11.2° + 17.4 = 28.6°$

(b) $R = 120 \text{k}\Omega$.

$|Z| = \sqrt{R^2 + X_c^2} = 124142 \Omega$
and

$\phi = \tan^{-1} (X_c/R) = 14.84°$

$i_c = (E_m/|Z|) \sin(\omega t + \phi)$

$\quad = (2.74 \times 10^{-3}) \sin (a + 14.84°)$

The capacitor voltage

$v_c = (E_m X_c/|Z|) \sin(\omega t + \phi - 90°)$

$\quad = 87.13 \sin (a - 75.2°) = 63$

where 63 V is the Diac break-over voltage. Hence

$\sin(a - 75.2°) = 0.723$

$(a - 75.2°) = \sin^{-1} (0.723) = 46.3°$

$a = 46.3° + 75.2° = 121.5°$

3.4   SINGLE-PHASE FULL-WAVE CONTROLLED RECTIFIER

In the half-wave controlled rectifier, full use is not being made of the a.c. supply voltage waveform; only the positive half-cycle is used. With the full-wave controlled rectifier, both positive and negative half-cycles are used. There are a number of circuit configurations that can be used to achieve full-wave control. Circuits shown in Figs 3.5–3.9 use one, two and four switching devices.

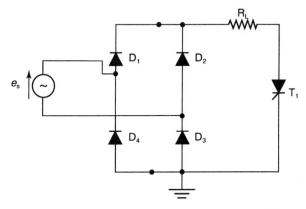

Figure 3.5   Single thyristor

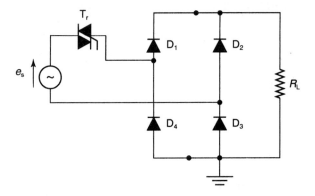

Figure 3.6   Single Triac

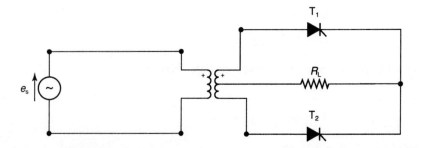

Figure 3.7   Two thyristors

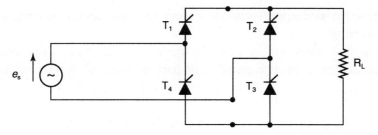

Figure 3.8   Four thyristors

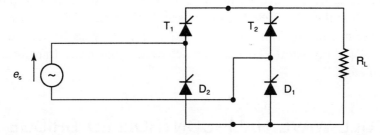

Figure 3.9   Two thyristors and two diodes

3.5   ONE- TO FOUR-QUADRANT OPERATION

Full-wave controlled rectifiers with resistive loads, or with inductive loads with free-wheel diodes, act in what is known as the first quadrant. This means that load voltage and current act in the positive direction only. Figure 3.10 shows how the four quadrants are defined.

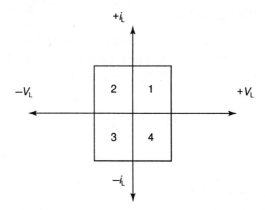

Figure 3.10

One-quadrant operation is typical of the circuits in Figs 3.5, 3.6, and 3.9. Two-quadrant operation is possible with the circuits in Figs 3.7 and 3.8, if the load is inductive or contains a d.c. motor. In the second quadrant, the load voltage reverses and energy can flow from the load to the source, with the system inverting.

Four-quadrant operation is possible using two full-wave fully controlled bridges connected in inverse parallel, or back-to-back. This type of operation occurs in a separately excited d.c. motor drive providing forward and reverse, speed and braking control.

Gating, or firing, of the switches requires the gate pulses to be synchronized to the a.c. mains voltage with a controllable delay on voltage zero. A simple circuit is the *R–C*–Diac combination, but this has a limited range of firing angle delay. Where a wide range of delay is required, a commercially produced module, such as the TDA2086A, can be used. In the case of the full-wave fully controlled bridge in Fig. 3.8, isolation is required of the gate drive circuits of at least two of the thyristors. This can be achieved using pulse transformers, or optical isolators.

3.6 FULL-WAVE HALF-CONTROLLED BRIDGE WITH RESISTIVE LOAD

Of the full-wave configurations, the half-controlled is the easiest to implement, since the two thyristors can be arranged to have a common cathode. The firing circuit can have a common train of pulses and only the forward-biased device will switch on at the arrival of a pulse on the two gates. It is still necessary to keep the mains neutral separate from the firing circuit common connection. The circuit arrangement and voltage waveforms are shown in Fig. 3.11.

Average load voltage (V_{av})

As for the half-wave case, average load voltage is found by calculating the area under the voltage curve and then dividing by the length of the base. For any delay angle a, the average load voltage is given by

$$V_{av} = (1/\pi) \int_{a}^{\pi} E_m \sin \theta \, \delta\theta = (E_m/\pi) \left[-\cos\theta\right]_{a}^{\pi}$$

Hence

$$\boxed{V_{av} = (E_m/\pi) \, (1 + \cos a)} \tag{3.5}$$

$$\boxed{I_{av} = V_{av}/R} \tag{3.6}$$

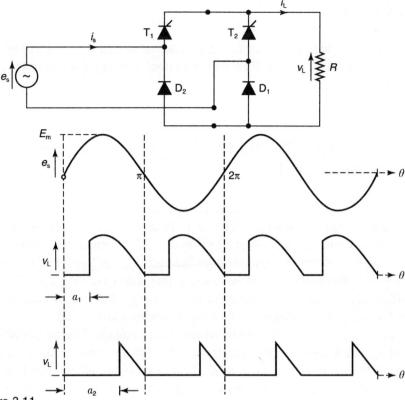

Figure 3.11

Root-mean-square load voltage (V_{rms})

The square root of the average value of the square of the time-varying voltage give the rms value:

$$V_{rms} = \sqrt{(1/\pi) \int_\alpha^\pi (E_m \sin \theta)^2 \, \delta\theta} \tag{3.7}$$

Using the identity that $\sin^2\theta = 0.5(1 - \cos2\theta)$

$$V_{rms} = E_m \sqrt{(1/\pi) \int_\alpha^\pi 0.5(1 - \cos 2\theta) \, \delta\theta}$$

$$= (E_m/\sqrt{2}) \sqrt{(1/\pi) [\theta - (\sin 2 \theta)/2)]_\alpha^\pi}$$

$$= (E_m/\sqrt{2}) \sqrt{(1/\pi) (\pi - \alpha + (\sin 2\alpha)/2)}$$

Now $(E_m/\sqrt{2}) = E_s$, the rms value of the supply voltage; hence

$$\boxed{V_{rms} = (E_s) \sqrt{(1 - (\alpha/\pi) + (\sin 2\alpha)/2\pi)}} \tag{3.8}$$

$$I_{rms} = V_{rms}/R \tag{3.9}$$

Compared to the half-wave circuit, rms values of current and voltage are both increased by a factor of $\sqrt{2}$, load power is doubled, and the converter power factor is improved.

3.7 HALF-CONTROLLED BRIDGE WITH HIGHLY INDUCTIVE LOAD

In the circuit in Fig. 3.11, the load is replaced by a large inductance. The assumption is that the load inductance is high enough to cause continuous steady load current. In the positive half-cycle, T_1 is turned on at delay angle a, and current flows to the load through the path T_1, load and D_1. The supply voltage passes through zero and reverses; if this was a resistive load T_1 would turn-off. However, due to the inductive stored energy, the load voltage reverses in order to keep the load current flowing, D_2 is forward-biased and conducts, and clamps the bottom of the load to virtually zero voltage. Energy stored in the load inductance keeps load current flowing through the path of D_2, T_1 and the load. At delay angle $\pi + a$, T_2 is fired, T_1 is reverse-biased and turns off, and load current flows through T_2, load and D_2. Once again, the supply voltage passes through zero, and load inductive energy forward biases D_1 to keep load current flowing. T_1 is then fired, D_2 turns off and the cycle is repeated.

The waveforms are shown in Fig. 3.12.

Average load voltage

As for resistive load

$V_{av} = (E_m/\pi) (1 + \cos a)$,

$I_{av} = V_{av}/R$

Root-mean-square supply current is given by

$$I_{rms} = \sqrt{(1/\pi) \int_{a}^{\pi} I_{av}^2 \delta\theta}$$
$$= I_{av} \sqrt{(\pi - a)/\pi} \tag{3.10}$$

Example 3.4 _____

A full-wave half-controlled bridge has a supply voltage of 220 V at 50 Hz. The firing angle delay $a = 90°$. Determine the values of average and rms currents,

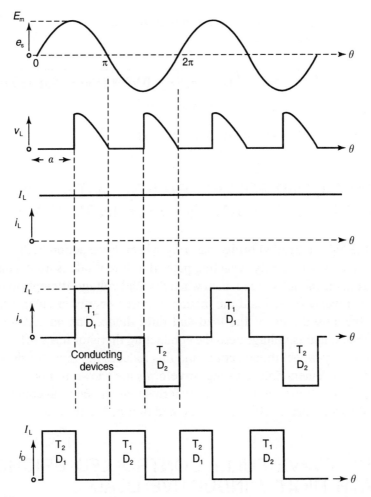

Figure 3.12

load power and power factor for (a) a resistive load of $R = 100\Omega$, (b) a highly inductive load with a resistance of 100Ω.

Solution

(a) V_{av} $= (E_m/\pi)(1 + \cos a) = (\sqrt{2} \times 220/\pi)(1 + \cos 90°) = 99\,V$

I_{av} $= V_{av}/R = 99/100 = 0.99\,A$

V_{rms} $= (E_s)\sqrt{(1 - (a/\pi) + (\sin 2a)/2\pi)}$

$= 220 \sqrt{(1 - (90/180) + \sin 180°/2\pi)}$

$= 155.6\,V$

$I_{rms} = V_{rms}/R$ $= 155.6/100 = 1.556\,A$

$P = I_{rms}{}^2 R$ $= (1.556)^2\, 100 = 242\,W$

$\cos \phi = P/E_{rms}\, I_{rms} = 242/220 \times 1.556 = 0.643$

(b) I_{av} $\qquad = 0.99\,\text{A (as in resistive load)}$

$\quad\ I_{rms} \qquad\qquad = \sqrt{(1/\pi)\int\limits_{a}^{\pi} I_{av}{}^2\,\delta\theta}$

$\qquad\qquad\qquad\qquad = I_{av}\sqrt{(\pi - a)/\pi} = 0.99\sqrt{(180 - 90)/180} = 0.7\,\text{A}$

$\quad P = I_{av}{}^2\,R \qquad\ = 0.99^2 \times 100 = 98\,\text{W}$

$\quad \cos\phi = P/E_{rms}\,I_{rms} = 98/220 \times 0.7 = 0.636$

3.8 HALF-CONTROLLED BRIDGE WITH FLY-WHEEL DIODE AND HIGHLY INDUCTIVE LOAD

Although the half-controlled bridge has a fly-wheel diode action built in, it uses one of the thyristors in the fly-wheeling path. If a third diode is used, connected directly across the inductive load, then when the load voltage attempts to reverse, this diode is reverse-biased and the inductive stored energy circulates the load current in the closed path of the load and third diode. The advantage of this method is that at mains voltage zero the conducting thyristor turns off instead of hanging on for fly-wheel diode action, and this reduces the thyristor duty cycle. The circuit arrangement and resulting waveforms are shown in Fig. 3.13.

It is clear from observation of the waveforms that values of average and rms voltage and current are unaffected by the addition of the third diode.

3.9 FULL-WAVE FULLY CONTROLLED BRIDGE WITH HIGHLY INDUCTIVE LOAD

The bridge thyristors can only conduct in one direction, but without diodes in the bridge the load voltage can reverse due to the load inductance, the load current continues to circulate, and current is circulated back to the mains against the direction of the mains voltage. In fact, the stored energy in the load is regenerating back to the supply. This is two-quadrant operation. The circuit and waveforms are given in Fig. 3.14.

In this circuit, the outgoing thyristors are turned off, or commutated, by reverse bias from the supply, when the incoming thyristors are switched on, i.e. there is no load current zero:

$$V_{av} = (1/\pi)\int\limits_{a}^{(\pi + a)} E_m \sin\theta\,\delta\theta$$

$$\quad = (E_m/\pi)\,[-\cos\theta]_{a}^{(\pi + a)}$$

$$\quad = (E_m/\pi)\,(\cos a + \cos a)$$

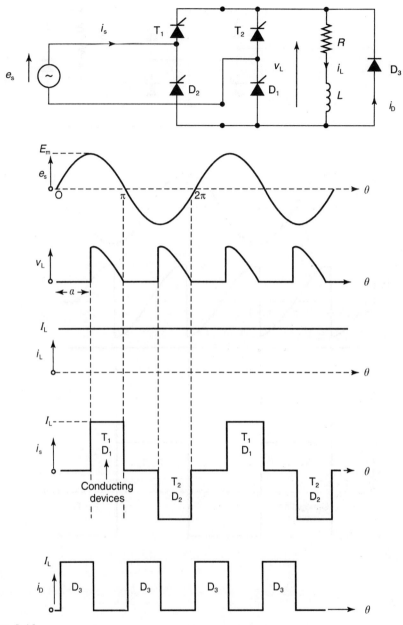

Figure 3.13

Hence

$$\boxed{V_{av} = (2E_m/\pi) \cos \alpha}$$ (3.11)

$I_{av} = V_{av}/R$

(This is assumed constant due to high inductance; thus $I_{av} = I_{rms}$.)

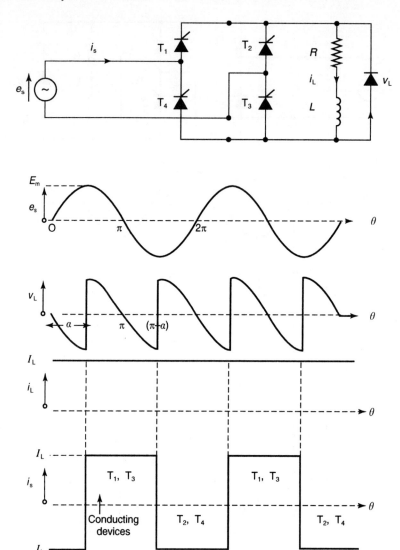

Figure 3.14

Load power,

$$P_L = I_{rms}^2 R$$

Overall power factor,

$$\cos \phi = P_L/E_{rms} I_{rms}$$

Note here that the load power factor is directly proportional to $\cos \alpha$, i.e.

$$\cos \phi = (I_{rms}^2 R)/E_{rms} I_{rms}$$

$$= (I_{rms} R)/E_{rms} = V_{av}/E_{rms}$$

$$= 2 (\sqrt{2}) E_{rms}\cos \alpha/\pi E_{rms}$$

$$\boxed{\cos \phi = 0.9 \cos \alpha} \qquad (3.12)$$

Example 3.5

A full-wave fully controlled bridge has a highly inductive load with a resistance of 55Ω, and a supply of $110\,\text{V}$ at $50\,\text{Hz}$.

(a) Calculate the values of load current, power and converter power factor for a firing angle delay $\alpha = 75°$.

(b) Sketch a curve showing the variation of average load voltage with firing angle delay.

Solution

(a) $\quad V_{av} \quad = (2E_m/\pi) \cos \alpha = (2\,\sqrt{2}.110/\pi) \cos 75°$

$\qquad\qquad = 99 \cos 75° = 25.6\,\text{V}$

$\quad I_{av} \quad = V_{av}/R = 25.6/55 = 0.446\,\text{A} = I_{rms}$

$\quad P_L \quad = I_{rms}^2 R = (0.446)^2\, 55 = 10.9\,\text{W}$

$\quad \cos \phi = 0.9 \cos \alpha = 0.9 \cos 75° = 0.233$

(b) The sketch is shown in Fig. 3.15.

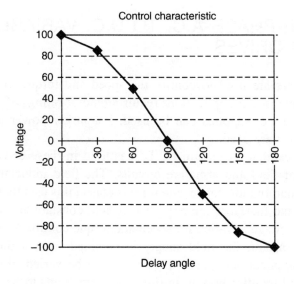

Figure 3.15

Example 3.6

A 240 V, 50 Hz supply feeds a highly inductive load of 50Ω resistance through a thyristor bridge that is (a) half-controlled, (b) fully-controlled. Calculate load current, power and power factor for each case when the firing angle delay $\alpha = 45°$.

Solution

(a) V_{av} $= (E_m/\pi)(1 + \cos\alpha)$

$\quad\quad\quad = (\sqrt{2} \times 240/\pi)(1 + \cos 45°) = 184.4\,\text{V}$

$\quad I_{av}$ $= V_{av}/R = 184.4/50 = 3.69\,\text{A}$

$\quad I_{rms}$ $= I_{av}\sqrt{(\pi - \alpha)/\pi} = 3.69\sqrt{(180 - 45)/180} = 3.2\,\text{A}$

$\quad P$ $= 3.2^2 \times 50 = 512\,\text{W}$

$\quad \cos\phi = P/E_s\,I_{rms} = 512/240 \times 3.2 = 0.667$

(b) V_{av} $= (2E_m/\pi)\cos\alpha$

$\quad\quad\quad = (2 \times 339/\pi)\cos 45° = 152.6\,\text{V}$

$\quad I_{av}$ $= V_{av}/R = 152.6/50 = 3.05\,\text{A}$

$\quad P_L$ $= I_{rms}^2\,R = 3.04^2 \times 50 = 466\,\text{W}$

$\quad \cos\phi = 0.9\cos\alpha = 0.9\cos 45° = 0.636$

3.10 SINGLE-PHASE A.C. TO D.C. VARIABLE SPEED DRIVES

Thyristor converters are used to control the speed and torque of d.c. motors driven from a.c. mains. For applications requiring motor ratings of about 10kW or so, single-phase a.c. supplies can be used. For larger power applications, three-phase supplies are used.

The separately excited d.c. motor (SEDC) is used in these drives, allowing separate control of field and armature circuits. The field inductance is much larger than the armature inductance, and for this reason it is usually the case that the field current, and therefore the pole flux, is held constant at its rated value and the armature voltage is varied for speed and torque control. Where the field current is held constant, the field supply is obtained from a diode rectifier bridge. In some applications, the field current may be varied; this requires a thyristor-controlled rectifier bridge. In this case, care needs to be exercised in using the torque constant, since the magnetization curve, pole flux against field current, is non-linear.

The armature is supplied from a thyristor-controlled rectifier bridge, which can be half-controlled if only forward control of speed and torque is required. It must be fully controlled if regeneration is required, and there must be two fully controlled full-wave bridges connected in inverse parallel if forward and reverse control of speed and torque are required.

3.11 THE SEDC MOTOR

The circuit diagram of the motor is shown in Fig. 3.16. For steady-state operation, the following performance equations can be used to analyse the drives behaviour:

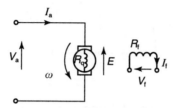

Figure 3.16

- field current:

$$V_f = I_f R_f \qquad (3.13)$$

- armature current:

$$V_a = I_a R_a + E \qquad (3.14)$$

- generated voltage:

$$E = k_v \omega \qquad (3.15)$$

ω is the armature speed in rad/s and k_v is the armature voltage constant in V/rad·s.

Equating electrical power in the armature to developed power on the motor shaft:

$$E I_a = \omega T \text{ (W)}$$

Torque

$$T = E I_a / \omega = k_v I_a \quad \text{(Nm)} \qquad (3.16)$$

Thus k_v is also the motor torque constant in Nm/A.

3.12 FULLY CONTROLLED BRIDGE WITH SEDC MOTOR

Assume a steady armature load resulting in steady armature speed ω and generated voltage E (see Fig. 3.17). Also assume that motor inductances are large enough to give a steady armature current between the firing periods.

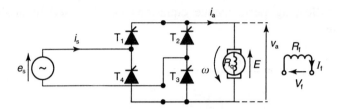

Figure 3.17

There is now a limited range of firing angle delay. The bridge thyristors will not turn on until they are foward-biased, i.e. until $v_s > E$. This limits the firing range to between a_1 and a_2, as shown in the waveforms in Fig. 3.18.

T_1 and T_3 are turned on in the positive half-cycle, in this case with a delay angle of about 30°. The conducting thyristors do not turn off at mains voltage zero because the motor inductance acts to keep the current flowing. Turn-off occurs at $(\pi + a)$ in the cycle when T_2 and T_4 are switched on.

The drive performance is determined using the following equations:

$E = k_v \omega$ (V)   assumed constant

As for highly inductive load,

$V_{av} = (2E_m/\pi) \cos a$ (V)

Average armature current,

$I_{av} = (V_{av} - E)/R_a$ (A)

Motor torque,

$T = k_v I_{av}$ (Nm)

Example 3.7 _____

A separately excited d.c. motor is driven from a 240V, 50Hz supply using a fully controlled thyristor bridge. The motor has an armature resistance R_a of 1.0Ω, and an armature voltage constant k_v of 0.8 V/rad·s. The field current is constant at its rated value. Assume that the armature current is steady.

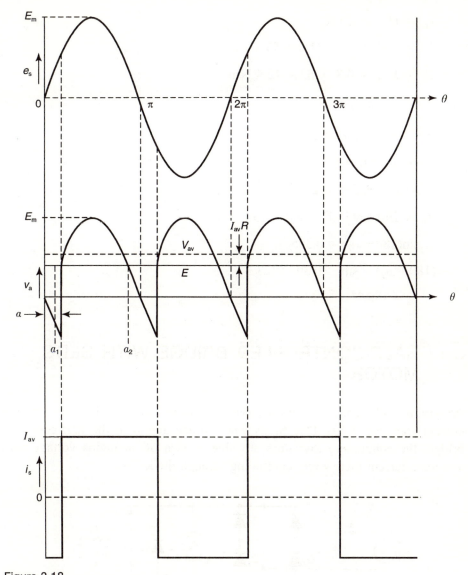

Figure 3.18

(a) Determine the values of armature current and torque for an armature speed of 1600 rev/min and a firing angle delay of 30°.

(b) Calculate the limits of the firing angle delay for this speed.

Solution

(a) $E = k_v \omega = (0.8 \times 1600 \times 2\pi/60) = 134\,\text{V}$

   $V_{av} = (2E_m/\pi) \cos a$

   $= (2\sqrt{2} \times 240/\pi) \cos 30° = 187.1\,\text{V}$

$$I_{av} = (V_{av} - E)/R_a$$

$$= (187.1 - 134)/1.0 = 53 \text{ A}$$

$$T = k_v I_{av} = 0.8 \times 53 = 42.4 \text{ Nm}$$

(b) $\sqrt{2} \times E_s \sin \theta \geq E$

$\sqrt{2} \times 240 \sin \alpha_1 = 134$

$\therefore \sin \alpha_1 = 0.395$

$\alpha_1 = \sin^{-1} 0.395 = 23.3°$

$\sqrt{2} \times 240 \sin(180° - \alpha_2) = 134$

$\therefore \sin(180 - \alpha_2) = 0.395$

$(180 - \alpha_2) = \sin^{-1} 0.395 = 23.3°$

$\therefore \alpha_2 = 156.7°$

3.13 HALF-CONTROLLED BRIDGE WITH SEDC MOTOR

As with the fully controlled bridge assume constant speed and steady armature current (see Fig. 3.19). Due to fly-wheel diode action in the half-controlled bridge, the conducting thyristors are able to turn off at mains voltage zero, armature current meanwhile continuing through diode D_3.

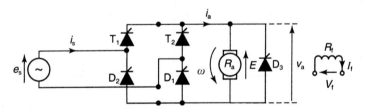

Figure 3.19

Unlike the fully controlled bridge, which has two-quadrant operation, the half-controlled bridge only allows one-quadrant operation, and no regeneration is possible. Waveforms showing the operation of the circuit are given in Fig. 3.20. Limitations of firing angle delay to between α_1 and α_2 also apply to this bridge.

Due to the absence of negative armature voltage, the average voltage V_{av} is higher than the fully controlled case and is the same expression as for average resistive load voltage, i.e.

$$V_{av} = (\sqrt{2} E_s/\pi) (1 + \cos a)$$

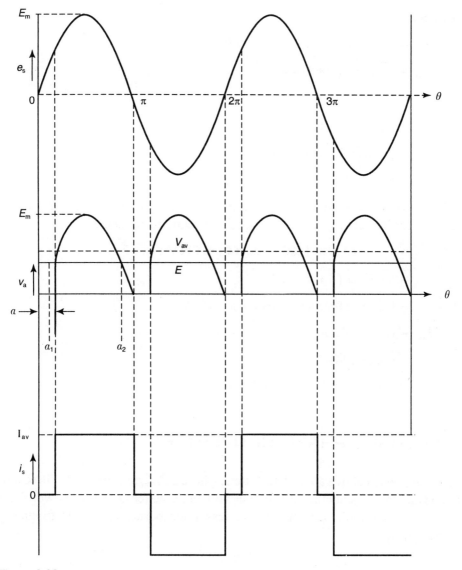

Figure 3.20

Other calculations of performance follow the same pattern as the fully controlled case.

Example 3.8 _____

A separately excited d.c. motor is driven from a 240V, 50Hz supply using a half-controlled thyristor bridge with a fly-wheel diode connected across the armature. The motor has an armature resistance R_a of 1.0Ω, and an armature voltage constant k_v of 0.8 V/rad/s. The field current is constant at its rated value. Assume that the armature current is steady.

Determine the values of armature current and torque for an armature speed of 1600 rev/min and a firing angle delay of (a) 30°, (b) 60°.

Solution

(a) $E = k_v \omega = (0.8 \times 1600 \times 2\pi/60) = 134\,\text{V}$

$V_{av} = \sqrt{2}\, E_s/\pi\, (1 + \cos a)$

$\qquad = (\sqrt{2} \times 240/\pi)(1 + \cos 30°) = 201.5\,\text{V}$

$I_{av} = (V_{av} - E)/R_a$

$\qquad = (201.5 - 134)/1.0 = 67.5\,\text{A}$

$T = k_v I_{av} = 0.8 \times 67.5 = 54\,\text{Nm}$

Note average armature voltage, current and torque are all higher than the fully controlled case (Example 3.7), for the same firing angle.

(b) $E = k_v \omega = (0.8 \times 1600 \times 2\pi/60) = 134\,\text{V}$

$V_{av} = \sqrt{2}\, E_s/\pi\, (1 + \cos a)$

$\qquad = (\sqrt{2} \times 240/\pi)(1 + \cos 60°) = 162\,\text{V}$

$I_{av} = (V_{av} - E)/R_a$

$\qquad = (162 - 134)/1.0 = 28\,\text{A}$

$T = k_v I_{av} = 0.8 \times 28 = 22.4\,\text{Nm}$

3.14 THREE-PHASE CONVERTERS

For load powers above about 10kW three-phase converters are used. These have the advantage of smoother d.c. output voltage, a better converter power factor and lower harmonic generation, i.e. less mainsborne, and radio-frequency, interference.

3.15 THREE-PHASE HALF-WAVE CONVERTER

A source of balanced three-phase star-connected voltages is shown in Fig. 3.21. E_{an}, E_{bn} and E_{cn} are the values of the rms phase voltages. Instantaneous phase voltages are given by the expressions:

$e_{an} = E_{pm} \sin \omega t$

$e_{bn} = E_{pm} \sin (\omega t - 120°)$

$e_{cn} = E_{pm} \sin (\omega t - 240°) = E_{pm} \sin (\omega t + 120°)$

where E_{pm} is the maximum value of the phase voltage. The instantaneous values of the line voltages are given by

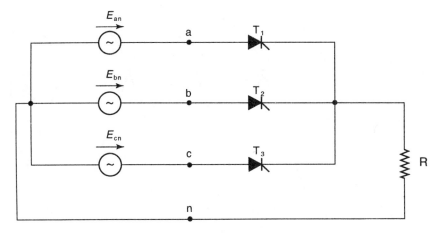

Figure 3.21

$$e_{ab} = (e_{an} - e_{bn}) = \sqrt{3}\, E_{pm} \sin(\omega t + 30°) = E_{lm} \sin(\omega t + 30°)$$

$$e_{bc} = (e_{bn} - e_{cn}) = \sqrt{3}\, E_{pm} \sin(\omega t - 90°) = E_{lm} \sin(\omega t - 90°)$$

$$e_{ca} = (e_{cn} - e_{an}) = \sqrt{3}\, E_{pm} \sin(\omega t - 210°) = E_{lm} \sin(\omega t - 210°)$$

where E_{lm} is the maximum value of the line voltage.

The waveforms of the line and phase voltages are given in Figs 3.22 and 3.23.

The half-controlled converter thyristors switch the phase voltages. Remember the thyristor will only turn on when a gate pulse is received if the anode is positive with respect to the cathode. For the circuit of Fig. 3.21, if T_3 was conducting then T_1 could be turned on just after the cross-over point of voltages e_{cn} and e_{an}, at angle 30° or $\pi/6$ rad. The phase voltages are then $0.5E_{pm}$. When T_1 turns on, T_3 is reverse-biased by the phase voltage e_{an} and turns off. This process is called line commutation. The next cross-over point is at angle 150° or $5\pi/6$ rad. T_2 is turned on here, commutating T_1.

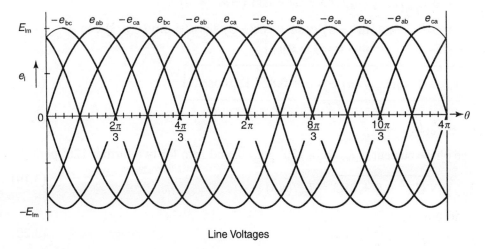

Line Voltages

Figure 3.22

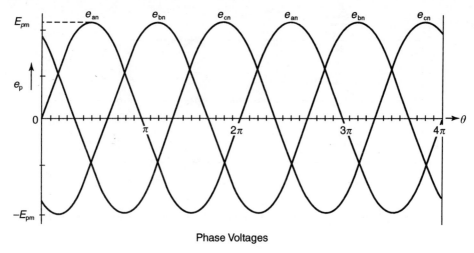

Phase Voltages

Figure 3.23

Load voltage and current waveforms are shown in Fig. 3.24 for a highly inductive load with steady current between the firing periods. The thyristors are turned on with delay angle a after the phase voltage cross-over points.

Average load voltage

The expression for the average load voltage is derived from the waveform of the load voltages in Fig. 3.25.

Each firing period is symmetrical, and taking the reference point as phase voltage maximum, the average load voltage is found as follows:

$$V_{av} = E_{pm}/(2\pi/3) \int_{-(\pi/3)+a}^{(\pi/3)+a} \cos\theta \ \delta\theta = 3 \ E_{pm}/2\pi \ [\sin\theta]$$

$$= 3 \ E_{pm}/2\pi \ \{ \sin(\pi/3 + a) - \sin(-\pi/3 + a) \}$$

$$= 3 \ E_{1pm}/\pi \ \{ \sin(\pi/3) \cos a + \cos(\pi/3) \sin a - \sin(-\pi/3) \cos a - \cos(-\pi/3) \sin a\}$$

$$\boxed{V_{av} = 3 \ \sqrt{3}(E_{pm}/2\pi) \cos a = 0.827 \ E_{pm}\cos a} \tag{3.17}$$

$$I_{av} = V_{av}/R \tag{3.18}$$

The phase current is I_{av} for a $120°$ period, and zero for two further $120°$ periods:

$$I_{rms} = \sqrt{(I_{av}^2 + 0 + 0)/3} = I_{av}/(\sqrt{3}) \tag{3.19}$$

Load power,

$$P = I_{av}^2 R \tag{3.20}$$

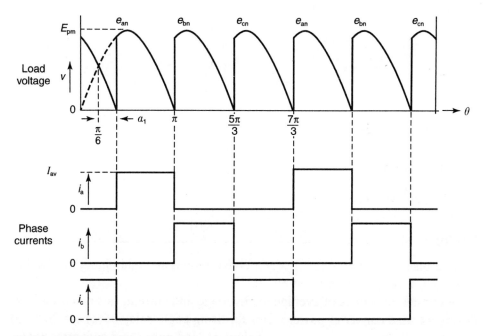

(a) Firing angle delay a_1

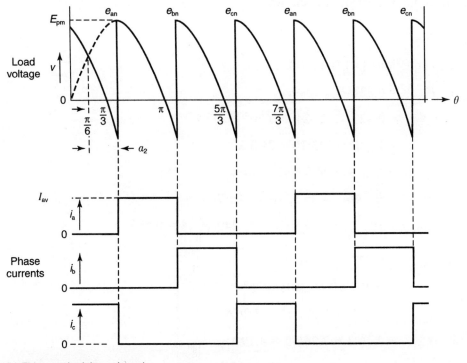

(b) Firing angle delay a_2 $(a_2 > a_1)$

Figure 3.24

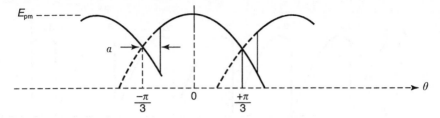

Figure 3.25

Converter power factor,

$$\cos\phi = P/3\, E_{rms}\, I_{rms} \tag{3.21}$$

Example 3.9

A three-phase half-controlled thyristor converter has an highly inductive load of 10 Ω, and a supply of 240 V at 50 Hz.

Determine the values of average load voltage and current, rms phase current, load power and converter power factor for firing angle delay of (a) $a = 30\,°$, (b) $a = 75°$. (c) What are the maximum values of load power and converter power factor obtainable from the circuit?

Solution

Using equations (3.17)–(3.21).

(a) V_{av} $= 0.827\, E_{pm} \cos a$

$= 0.827\, \sqrt{2} \times 240\cos 30° = 243\,V$

I_{av} $= V_{av}/R = 243/10 = 24.3\,A$

I_{rms} $= I_{av}/\sqrt{3} = 24.3\,/\,\sqrt{3} = 14\,A$

P $= I_{av}^2 R = 24.3^2 \times 10 = 5.9\,kW$

$\cos \phi = P/3\, E_p\, I_{rms} = 5900/3 \times 240 \times 14 = 0.585$

(b) V_{av} $= 0.827\, E_{pm} \cos a$

$= 0.827\, \sqrt{2} \times 240\cos 75° = 72.6\,V$

I_{av} $= V_{av}/R = 72.6\,/\,10 = 7.26\,A$

I_{rms} $= I_{av}/\sqrt{3} = 7.26/\sqrt{3} = 4.19\,A$

P $= I_{av}^2 R = 7.26^2 \times 10 = 527\,W$

$\cos \phi = P/3\, E_p\, I_{rms} = 527/3 \times 240 \times 4.19 = 0.175$

(c) V_{av} = 0.827 E_{pm} cos α = 0.827 $\sqrt{2} \times$ 240cos 0° = 281 V

I_{av} = V_{av}/R = 281 / 10 = 28.1 A

I_{rms} = $I_{av}/\sqrt{3}$ = 28.1 / $\sqrt{3}$ = 16.2 A

P = $I_{av}^2 R$ = 28.1$^2 \times$ 10 = 7.9 kW

cos ϕ = $P/3$ E_p I_{rms} = 7900/3 $\times$ 240 $\times$ 16.2 = 0.68

Assuming that the a.c. supply to the converter is obtained from the secondary windings of a three-phase transformer, then straight star connection as shown in Fig. 3.21 would not be suitable due to the d.c. component of the phase currents producing d.c. magnetizing ampere-turns in the transformer. This problem is overcome by using a zig-zag connected secondary winding which cancels out the d.c. ampere-turns in each phase (Lander, 1993, p. 57).

3.16 THREE-PHASE FULL-WAVE CONVERTER

This converter can control loads of up to about 100kW. If two-quadrant operation is required a fully controlled three-phase bridge is required. If only one-quadrant operation will do, then a half-controlled three-phase bridge can be used, with its simpler firing circuits. The circuit in Fig. 3.26 shows a fully controlled full-wave three-phase thyristor bridge converter.

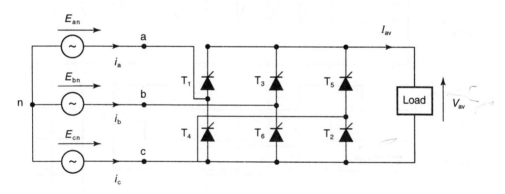

Figure 3.26

In this circuit, two thyristors are fired simultaneously. The reference point for successful turn-on is now the cross-over of the line voltages. Firing pulses are required every 60°, or $\pi/3$ rad. Thyristors T_1, T_3 and T_5 are referred to as the positive group; thyristors T_4, T_6 and T_2 are the negative group. The firing sequence of the thyristors is tabulated below for 60° intervals. Each thyristor is conducting for a 120° period, and is off for a 240° period:

Positive group	T_1	T_1	T_3	T_3	T_5	T_5	T_1
Negative group	T_6	T_2	T_2	T_4	T_4	T_6	T_6

Figure 3.27 shows the waveforms of line voltages, load voltage, thyristor currents and one line current for a firing angle delay a of about 30°. The line current waveform for i_b is the same shape as i_a but lags by 120°and i_c is the same shape but lags by 240°.

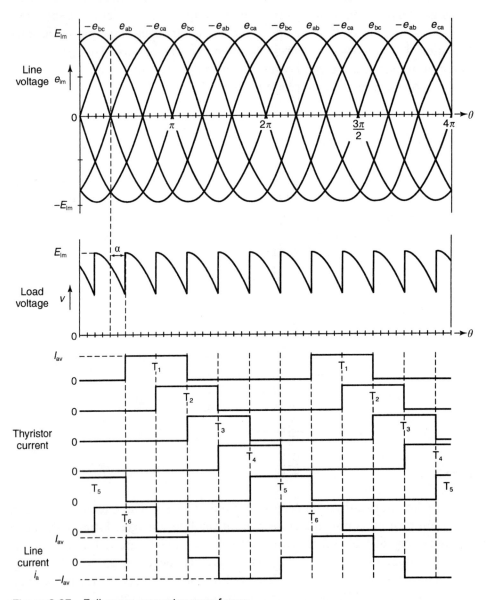

Figure 3.27　Full-wave converter waveforms

Average load voltage (V_{av})

Each firing period is symmetrical, and taking the reference point as line voltage maximum, the average load voltage is found as in Fig. 3.28:

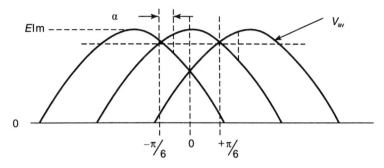

Figure 3.28

$$V_{av} = E_{lm}/(\pi/3) \int_{-(\pi/6)+\alpha}^{(\pi/6)+\alpha} \cos\theta \ \delta\theta = 3E_{lm}/\pi \ [\sin\theta]_{-\pi/6 \ + \ \alpha}^{\pi/6 \ + \ \alpha}$$

$$= 3 \ E_{lm}/\pi \ \{ \ \sin(\pi/6 + \alpha) - \sin(-\pi/6 + \alpha) \ \}$$

$$= 3 \ E_{lm}/\pi \ \{\sin(\pi/6) \cos\alpha + \cos(\pi/6) \sin\alpha - \sin(-\pi/6) \cos\alpha$$
$$-\cos(\pi/6) \sin\alpha\}$$

$$\boxed{V_{av} = 3(E_{lm}/\pi) \cos\alpha = 3 \ \sqrt{3} \ (E_{pm}/\pi) \cos\alpha} \qquad (3.22)$$

$$I_{av} = V_{av}/R$$
$$I_{rms} = \sqrt{(I_{av}^2 + (-I_{av})^2 + 0)/3} = (\sqrt{2/3}) \ I_{av}$$

Example 3.10

The three-phase half-controlled converter of Example 3.9 is replaced by a three-phase fully controlled thyristor converter. Load and supply remain unchanged, i.e. highly inductive load of 10Ω and a three-phase supply of 240 V at 50 Hz.

Determine the values of average load voltage and current, rms phase current, load power and converter power factor for a firing angle delay of (a) $\alpha = 30°$, (b) $\alpha = 75°$. (c) What are the maximum values of load power and converter power factor obtainable from the circuit?

Compare the solutions with those of Example 3.9.

Solution

(a)　$V_{av} = 3\sqrt{3}(E_{pm}/\pi)\cos\alpha$

$\qquad = 3\sqrt{3}\times(\sqrt{2}\times 240/\pi)\cos 30° = 486\,V$

$\quad I_{av} = V_{av}/R = 486/10 = 48.6\,A$

$\quad I_{rms} = \sqrt{(I_{av}^2 + (-I_{av})^2 + 0)/3} = (\sqrt{2/3})\,I_{av} = 39.7\,A$

$\quad P = V_{av}\,I_{av} = (48.6)^2\times 10 = 23.62\,kW$

$\quad \cos\phi = P/3E_{p(rms)}\,I_{rms} = 23.62\times 10^3/(3\times 240\times 39.7) = 0.826$

(b)　$V_{av} = 3\sqrt{3}(E_{pm}/\pi)\cos\alpha$

$\qquad = 3\sqrt{3}\times(\sqrt{2}\times 240/\pi)\cos 75° = 145\,V$

$\quad I_{av} = V_{av}/R = 145/10 = 14.5\,A$

$\quad I_{rms} = \sqrt{(I_{av}^2 + (-I_{av})^2 + 0)/3} = (\sqrt{2/3})\,I_{av} = 11.8\,A$

$\quad P = V_{av}\,I_{av} = (14.5)^2\times 10 = 2.1\,kW$

$\quad \cos\phi = P/3E_{p(rms)}\,I_{rms} = 2.1\times 10^3/(3\times 240\times 11.8) = 0.247$

(c)　$V_{av} = 3\sqrt{3}(E_{pm}/\pi)\cos\alpha$

$\qquad = 3\sqrt{3}\times(\sqrt{2}\times 240/\pi)\cos 0° = 561\,V$

$\quad I_{av} = V_{av}/R = 561/10 = 56.1\,A$

$\quad I_{rms} = \sqrt{(I_{av}^2 + (-I_{av})^2 + 0)/3} = (\sqrt{2/3})\,I_{av} = 45.8\,A$

$\quad P = V_{av}\,I_{av} = (56.1)^2\times 10 = 31.4\,kW$

$\quad \cos\phi = P/3E_{p(rms)}\,I_{rms} = 31.4\times 10^3/(3\times 240\times 45.8) = 0.954$

Comparison:

	$\alpha = 0°$	$\alpha = 0°$	$\alpha = 30°$	$\alpha = 30°$	$\alpha = 75°$	$\alpha = 75°$
	HC	FC	HC	FC	HC	FC
V_{av} (V)	281	561	243	486	72.6	145
I_{av} (A)	28.1	56.1	24.3	48.6	7.26	14.5
I_{rms} (A)	16.2	45.8	14	39.7	4.19	11.8
P (kW)	7.9	31.4	5.9	23.6	0.53	2.1
$\cos\phi$	0.68	0.954	0.585	0.826	0.175	0.247

The fully controlled bridge has twice the average load voltage and current, 2.83 times the rms current, four times the load power and a power factor improved by a factor of 1.4, compared with the half-controlled case.

3.17 THE *p*-PULSE CONVERTER

A general expression for the average load voltage of any fully controlled converter can be developed by reference to the waveform sketch in Fig. 3.29.

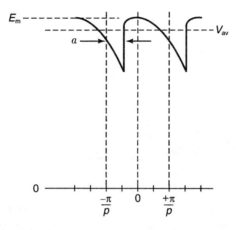

Figure 3.29

Average load voltage

$$V_{av} = E_m/(2\pi/p) \int\limits_{-\pi/p+a}^{+\pi/p+a} \cos\theta\ \delta\theta = p\ E_m/2\pi\ [\sin\theta]\limits_{-(\pi/p)+a}^{+(\pi/p)+a}$$

$$= p\ E_m/2\pi\ \{\ \sin(\pi/p + a) - \sin(-\pi/p + a)\ \}$$

$$= p\ E_m/2\pi\ \{\sin(\pi/p)\cos a + \cos(\pi/p)\sin a - \sin(-\pi/p)\cos a - \cos(\pi/p)\sin a\}$$

$$\boxed{V_{av} = (p\ E_m/\pi)\sin(\pi/p)\cos a} \qquad (3.23)$$

Now the average load voltages for the different converters are:

(a) Single-phase, $p = 2$:

$$V_{av} = (2E_m/\pi)\cos a$$

(b) Three-phase half-wave; $p = 3$:

$$V_{av} = (3\sqrt{3}\ E_{pm}/2\pi)\cos a$$

(c) Three-phase full-wave, $p = 6$,

$$V_{av} = (3\sqrt{3}\ E_{pm}/\pi)\cos a$$

3.18 TWELVE-PULSE CONVERTER

If higher output voltages and load power than those obtainable from the full-wave fully controlled bridge are required, then converters can be connected in series, as show in Fig. 3.30.

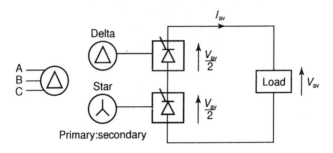

Figure 3.30

The two secondary windings have phase voltages 30° apart. In order for each secondary winding to produce the same voltage magnitude, the delta winding requires $\sqrt{3}$ more turns than the star winding. The easiest method of analysis is to consider each converter producing half the average load voltage. The average load voltage is then twice that produced by each six-pulse converter. Thus

$$V_{av} = 2 \times 3(E_{1m}/\pi) \cos \alpha$$

$$= 6(E_{1m}/\pi) \cos \alpha$$

The alternative method of analysis is to show a single load voltage waveform with an altered maximum line voltage to take into account the phasor addition of the two bridge outputs, as in Fig. 3.31. Now, we can see from the figure that

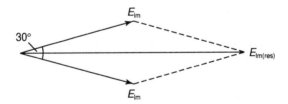

Figure 3.31

$$E_{1m(res)} = 2E_{1m} \cos 15° = 1.932 E_{1m}$$

The firing periods are now only 30° apart, i.e. 12 pulses per cycle, and the average load voltage is give by

$$V_{av(res)} = (p\, E_{1m(res)}/\pi) \sin \pi/p \cos \alpha = 12 \times 1.932\, E_{1m} (\sin \pi/12) \cos \alpha$$

$$= (6\, E_{1m}/\pi) \cos \alpha$$

3.19 SPEED REVERSAL AND REGENERATIVE BRAKING OF SEDC MOTOR DRIVES

As shown in equation (3.23), fully controlled converters have average output voltage give by

$$V_{av} = (p\,E_m/\pi)\sin(\pi/p)\cos a$$

This means that the increasing firing angle delay will reverse the output voltage at $a > 90°$ (see Fig. 3.32). The ability to reverse the converter voltage gives a means for speed reversal of the motor.

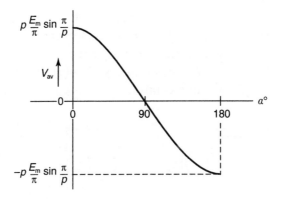

Figure 3.32

The d.c. separately excited motor requires either armature current or field current reversal to drive the motor in the reverse direction. In practice, field inductance is much larger than armature inductance and so field reversal is slower. The armature current can be reversed using either a contactor or two anti-parallel fully controlled bridges, as shown in Figs 3.33(a) and (b). Only one bridge is operated at any one time; the other bridge is inhibited.

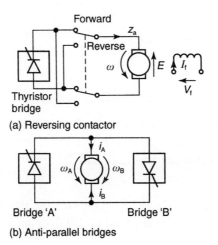

Figure 3.33

For reversal of speed and regenerative braking, the sequence of operations is shown in Figs 3.34(a)–(d). Assume the drive is in the forward motoring condition as shown in Fig. 3.34(a), and the requirement is to reverse the direction of rotation. The firing angle is increased to reduce the armature current to zero. The contactor can then be safely operated as indicated in Fig. 3.34(b). The firing angle is further increased until the generated voltage exceeds the converter voltage; regenerative braking comes into operation; energy is extracted from the armature and fed back to the supply; the motor brakes and generated voltage and speed both fall to zero. The armature voltage is brought back to the normal rectifying mode by phase angle control and speed builds up in the reverse direction to the required value, as shown in Fig. 3.34(c).

To return to the forward direction of speed, the firing angle is increased until armature current is again zero. The contactor is operated as in Fig. 3.34(d), generated voltage exceeds converter voltage, regenerative braking occurs, and armature speed falls to zero. As shown in Fig. 3.34(a), speed is built up in the forward direction by phase angle control.

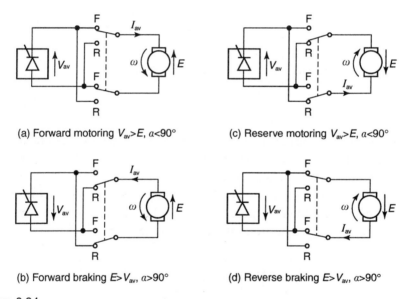

(a) Forward motoring $V_{av}>E$, $a<90°$ (c) Reserve motoring $V_{av}>E$, $a<90°$

(b) Forward braking $E>V_{av}$, $a>90°$ (d) Reverse braking $E>V_{av}$, $a>90°$

Figure 3.34

Example 3.11

A separately excited d.c. motor is driven from a three-phase fully controlled converter with a contactor for speed reversal. The line voltage is 440 V at 50 Hz. The motor armature resistance is $0.12\,\Omega$ and the armature voltage constant $k_v = 2.0$ V/rad/s at rated field current.

(a) Determine the firing angle delay necessary, at rated field current, for an armature current of 40 A at a speed of 1500 rev/min.

(b) The contactor is operated after delay angle control is used to bring the armature current to zero, at the start of the sequence for speed reversal. Assuming that initially the generated armature voltage remains constant, what delay angle is required to set the armature current back to 40 A for regenerative braking and what power is available for regeneration?

Solution

(a) $E = \omega k_v$

$$= (1500 \times 2\pi/60)2.0 = 314.2\,V$$

$V_{av} = E + I_{av}R_A$

$$= 314.2 + (40 \times 0.12) = 319\,V$$

$V_{av} = (3E_{1m}/\pi) \cos a$

$$= (3 \times 440\,\sqrt{2}/\pi) \cos a = 594 \cos a$$

$\therefore 594 \cos a = 319$

$\cos a = 319/594 = 0.537$

$a = 57.5°$

(b) $E - I_{av}R_a = -V_{av}$

$314.2 - (40 \times 0.12) = - V_{av} = -594 \cos a$

$594 \cos a = -309.4$

$\cos a = -0.52$

$a = 121.3°$

Regenerative braking power, P, is given by

$P = V_{av} I_{av} = 309.4 \times 40 = 12.4\text{kW}$

Braking torque $= P/\omega = 12400/(1500 \times 2\pi/60) = 79\,\text{Nm}$

3.20 SELF-ASSESSMENT TEST

1 What type of thyristor converter is required for operation in:

(a) the first quadrant

(b) the first and second quadrants

(c) all four quadrants?

2 An a.c. supply drives a half-controlled thyristor bridge with a resistive load. Sketch the circuit diagram, and the voltage and current waveforms.

3 A highly inductive load is driven from a fully controlled thyristor bridge connected to a.c. mains. Sketch the circuit diagram, and the voltage and current waveforms.

4 State the equations for average armature voltage, generated voltage, armature current and torque for a separately excited d.c. motor driven from a.c. mains via a fully controlled thyristor bridge. Why is the firing angle range limited in such a circuit?

5 State the equations for average armature voltage, generated voltage, armature current and torque for a separately excited d.c. motor driven from a.c. mains via a half-controlled thyristor bridge. Is the firing angle range limited in such a circuit?

6 Calculate the values of maximum output voltage obtainable from p-pulse converters with two, three, six and 12 pulses. Assume a phase input voltage of 250V, 50Hz in each case.

7 What type of converter would be required to drive a 100kW SEDC motor in forward and reverse directions, and to give regenerative braking? Briefly explain the sequence of events in the transition between forward and reverse motoring.

3.21 PROBLEMS

1 A resistive load is supplied with variable voltage d.c. from a Triac full-wave rectifier bridge combination connected to an a.c. supply. Draw the circuit diagram and, with the aid of voltage waveform sketches, calculate the load power dissipation for load resistance of 100Ω, a voltage supply of 110V at 50Hz, and firing angle delays of 45°, and (b) 135°. Assume ideal switching devices.

2 A full-wave fully controlled thyristor a.c. to d.c. converter supplies power to (a) a resistive load of 250Ω, (b) a highly inductive load with a resistance of 250Ω. The a.c. supply is 240V, 50Hz. Determine the values of load average and rms voltage and current, load power and converter power factor for (a) $a = 30°$, (b) $a = 60°$. Assume ideal switching devices.

3 Two of the bridge thyristors in Q.2 are replaced by diodes to make the converter half-controlled. Repeat the calculations of Q.2 for this converter and compare the performance of both converters.

4 A single-phase fully controlled SEDC motor drive has an armature voltage constant of 0.9 V/rad/s, and an armature resistance of 0.75Ω. The field current is held constant at its rated value. Mains supply to the drive is 250V, 50Hz.

Determine the average armature voltage, current and torque at an armature speed of 1200 rev/min, with a firing angle delay of (a) $\alpha = 30°$, (b) $\alpha = 70°$. Assume steady armature current.

5 Compare the performance of the drive of Q4 with that of the half-controlled case given for the same conditions, i.e. repeat the calculations in Q4 for a single-phase half-controlled SEDC drive.

6 A separately excited d.c. motor has an armature resistance of 1.0Ω and an armature voltage constant of 0.8 V/rad/s. Determine the average armature current and torque output, with firing angle delay of $\alpha = 45°$ and armature speed of 1600 rev/min. The armature is driven by:

(a) a full-wave fully controlled single-phase converter

(b) a half-controlled single-phase converter

(c) a three-phase half-wave converter

(d) a three-phase full-wave converter.

The supply phase voltage is 230V in each case.

7 A three-phase half-wave converter drives an SEDC motor rated at 500A, 1000 rev/min with $R_a = 0.02Ω$ and $k_v = 0.8$ Nm/A (V/rad/s). Find the delay angle required for an output torque of 400Nm at rated speed.

The line voltage input to the converter is 415V at 50Hz.

4 DC to AC inverters

The inverter provides a.c. load voltage from a d.c. voltage source. The semiconductor switches can be BJTs, thyristors, Mosfets, IGBTs etc. The choice of power switch will depend on rating requirements and ease with which the device can be turned on and off.

A single-phase inverter will contain two or four power switches arranged in half-bridge or full-bridge topologies. Half-bridges have the maximum a.c. voltage limited to half the value of the full d.c. source voltage and may need a centre tapped source. Full-bridges have the full d.c. source voltage as the maximum a.c. voltage. Where the d.c. source voltage is low, e.g. 12 V or 24 V, the voltage drop across the conducting power switches is significant and should be taken into account both in calculation and in selection of the switch.

The a.c. load voltage of the inverter is essentially a square wave, but pulse-width-modulation methods can be used to reduce the harmonics and produce a quasi-sine wave. If higher a.c. voltages than the d.c. source voltage are required, then the inverter will require a step-up transformer.

The output frequency of the inverter is controlled by the rate at which the switches are turned on and off, in other words by the pulse repetition frequency of the base, or gate, driver circuit.

Thyristors would only be used in very high power inverters, since on the source side there is no voltage zero, and a forced commutation circuit would be required to turn the thyristor off.

Some typical single-phase inverters are considered in the following sections. The switching device shown is a BJT, but could be any switch, the choice being determined by availability of required rating and ease of turn-on and turn-off.

Care must be taken not to have two switches 'on' together, shorting out the d.c. source. There must be either a dead-time between switches or an inhibit circuit to ensure this does not happen.

4.1 HALF-BRIDGE WITH RESISTIVE LOAD

In the circuit of Fig. 4.1, V_{p1} and V_{p2} are pulse generators acting as base drivers for the two transistors. V_{p1} provides a pulse every periodic time, T, of the chosen inverter frequency. V_{p2} provides a pulse every periodic time, but delayed by $T/2$ on V_{p1}. Figure 4.2 shows the shape of the output voltage wave form across the load resistor R_1.

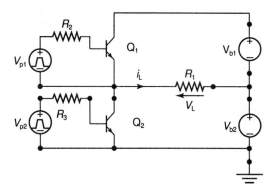

Figure 4.1

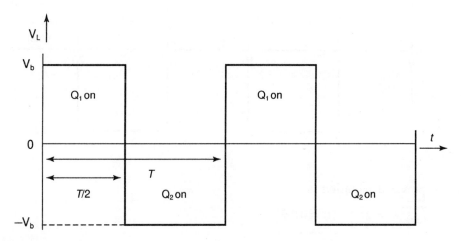

Figure 4.2

Inverter frequency, $f = 1/T$ (s).

Root mean square load voltage,

$$V_{rms} = \sqrt{(2/T) \int_0^{(T/2)} (V_b)^2 \, \delta t} = V_b$$

where $V_b = V_{b1} = V_{b2}$.

Example 4.1

A half-bridge inverter, as shown in Fig. 4.1, has $V_{b1} = V_{b2} = 20\,V$. The load is resistive with $R = 10\,\Omega$. Inverter frequency is $100\,Hz$. Sketch and scale the load current waveform and determine the load power dissipation.

Solution

Assume ideal switches.

Maximum load current = rms load current

$= V_{b1}/R = 20/10 = 2\,\text{A}$

Periodic time $= 1/f = 1/100 = 10\,\text{ms}$

The load current waveform is therefore as shown in Fig. 4.3.

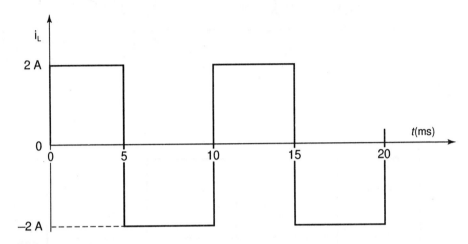

Figure 4.3

Load power dissipation is

$P_{L} = I_{rms}{}^{2}R = 2^{2} \times 10 = 40\,\text{W}$

4.2 HALF-BRIDGE INVERTER WITH RESISTIVE LOAD AND CAPACITIVE ELEMENTS

In the circuit in Fig. 4.4, capacitors replace the centre-tapped battery, and a single battery acts as the d.c. source. C_1 and C_2 are each charged to $V_b/2$ prior to switching. When Q_1 is switched on, C_1 capacitor voltage is applied across the load, load current flows and exponentially decays until Q_1 is switched off and Q_2 is switched on to connect C_2 as a reverse voltage across the load. Load current flows in reverse through the load, exponentially decaying until Q_2 turns off and Q_1 turns on to repeat the cycle. Figure 4.5 shows the typical load voltage waveform.

Example 4.2 _____

The circuit shown in Fig. 4.4 has $V_b = 40\,\text{V}$, $C_1 = C_2 = 200\,\mu\text{F}$, $R_1 = 10\,\Omega$ and an

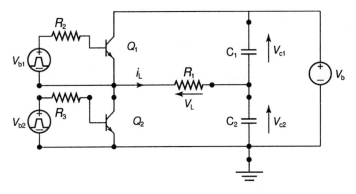

Figure 4.4

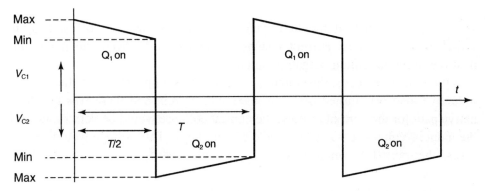

Figure 4.5

inverter frequency of 100 Hz. Assuming ideal switches, determine maximum and minimum values of load voltage, assuming the waveform is symmetrical.

Solution

$T = 1/f = 1/100 = 10\,\text{ms}$

$\tau = CR = 200 \times 10^{-6} \times 10 = 2\,\text{ms}$

For a symmetrical waveform

$$V_{c1(\text{max})} = V_{c2(\text{max})}$$

$$V_{c1(\text{min})} = V_{c2(\text{min})}$$

Also

$$V_{c1(\text{max})} = V_b - V_{c2(\text{min})}$$

and

$$V_{c2(\text{max})} = V_b - V_{c1(\text{min})}$$

Hence

$$V_{c1(max)} = 40 - V_{c1(max)} \exp(-t/CR)$$

$$= 40 - V_{c1(max)} \exp(-2.5)$$

$$\therefore V_{c1(max)} = 40/(1+ \exp(-2.5) = 37\,V$$

$$V_{c1(min)} = V_{c1(max)} \exp(-2.5) = 37 \times \exp(-2.5) = 3\,V$$

4.3   HALF-BRIDGE WITH PURELY INDUCTIVE LOAD

With resistive loads, at the half-cycle period Q_1 is switched off and Q_2 is switched on. The load current changes instantaneously from the maximum positive to the maximum negative value.

With inductive loads, instantaneous reversal is impossible. In fact, if an inductive load was turned off with maximum current and there was no alternative path for the current to flow, the current collapse would be very rapid, and the induced voltage, $e = L\,di/dt$, would be very large. This voltage would appear across the transistor switch in reverse and could destroy it.

To protect the transistor from this inductive surge, feedback diodes are connected in anti-parallel across the transistor to provide a path for the current to decay. In the circuit in Fig. 4.6, when Q_1 switches off at maximum positive current, the inductive voltage v_L reverses its polarity, the voltage rises above V_{b2} and forward biases D_2, allowing the decay of current until load current zero, when Q_2 will start the flow of current in the negative direction. When negative current reaches its maximum value, Q_2 switches off, v_L reverses, rising above V_{b1} and forward biasing diode D_1, until load current zero. Q1 will then allow current to grow in the positive direction and the cycle is repeated.

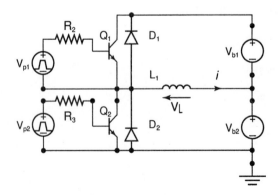

Figure 4.6

In Fig. 4.6, the current will ramp up and ramp down in response to switching. Initially a transient occurs until I_0, the inductor current at switching, is established. When the transient has disappeared, the waveform is symmetrical with I_0 and $-I_0$ as the initial conditions at switching.

The current waveform in Fig. 4.7 shows the conditions once the transient has disappeared. The circuit in the s domain at $t=0$ is given in Fig. 4.8.

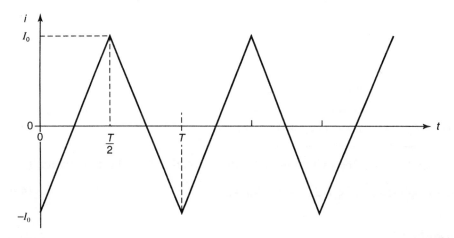

Figure 4.7

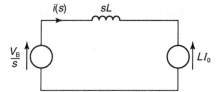

Figure 4.8

In this circuit

$i(s) = (\,(V_b/s) - LI_0)/sL$

$$\boxed{i(t) = (V_b/L)\, t - I_0} \tag{4.1}$$

At $t = T/2$, $i = +I_0$. Hence

$I_0 = ((V_b/L)(T/2)) - I_0$

$2I_0 = ((V_b/L)(T/2))$

$$\therefore \quad \boxed{I_0 = V_b T/4L = V_b/4fL} \tag{4.2}$$

Q_2 is now switched on with the initial condition given in equation (4.2). The s domain circuit is now as shown in Fig. 4.9.

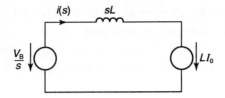

Figure 4.9

In this circuit

$i(s) = (LI_0 - (V_b/s))/sL$

$$i(t) = I_0 - (V_b/L)t = (V_bT/4L) - (V_b/L)t \qquad (4.3)$$

A further $t = T/2$ after switching, $i(t)$ will be equal to $-I_0$. From equation (4.3)

$$-I_0 = (V_bT/4L) - (V_b \ T/2L) = - \ V_b T/4L \qquad (4.4)$$

Example 4.2

A half-bridge inverter with a centre-tapped 40 V battery has a purely inductive load, L = 200 mH, and a frequency of 100 Hz. Determine the maximum load current and the load current at (a) t = 3 ms, (b) t = 5 ms, and (c) t = 8 ms, assuming that at t = 0 the transient had disappeared.

Solution

Periodic time is

$T = 1/f = 1/100 = 0.01\,s = 10ms$

Maximum curent is

$I_0 = V_b T/4L = 20 \times 10^{-2}/4 \times 0.2 = 250mA$

(a) $i(t) = (V_b/L)\, t - I_0$

$= (20/0.2)\, 3 \times 10^{-3} - 0.25 = 50mA$

(b) $i(t) = (V_b/L)\, t - I_0$

$= (20/0.2)\, 5 \times 10^{-3} - 0.25 = 250mA$

(c) At $t=5$ ms, the second switch turns on; 8 ms is 3 ms into the negative ramp. Thus

$i(t) = (V_b T/4L) - (V_b/L)t$

$= (250 \times 10^{-3}) - (20/0.2)\, 3 \times 10^{-3} = -50mA$

4.4   HALF-BRIDGE WITH AN *R–L* LOAD

The load is a series combination of resistance and inductance. The circuit is shown in Fig. 4.10.

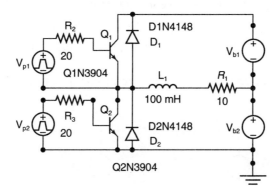

Figure 4.10

Assuming that I_0 is established, the transient has disappeared and load current waveform is symmetrical, the typical shape will be as shown in Fig. 4.11.

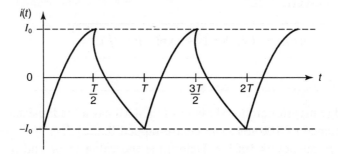

Figure 4.11

The *s*-domain circuit with Q_1 switched on is shown in Fig. 4.12.

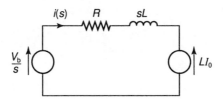

Figure 4.12

In this circuit

$i(s) = ((V_b/s) - LI_0)/(R + sL)$

$\quad = (V_b/s(R + sL)) - (LI_0 (R + sL))$

$\quad = (V_b/R) (R/s(s + R/L)) - LI_0/(R + sL)$

$$i(t) = (V_b/R)(1 - \exp(-Rt/L)) - I_0 \exp(-Rt/L) \qquad (4.5)$$

At the end of the half-cycle period, $t = T/2$ and $i(t) = I_0$; hence

$$I_0 = (V_b/R)(1 - \exp(-RT/2L)) - I_0 \exp(-RT/2L)$$

$$\therefore \quad I_0 = (V_b/R)(1 - \exp(-RT/2L)/(1 + \exp(-RT/2L)) \qquad (4.6)$$

Q_2 is now switched on and the s domain circuit changes to that shown in Fig. 4.13.

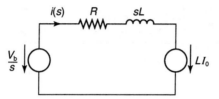

Figure 4.13

Now

$$i(s) = (LI_0 - V_b/s)/(R + sL)$$

$$i(t) = I_0 \exp(-Rt/L) - (V_b/R)(1 - \exp(-RT/2L)) \qquad (4.7)$$

Example 4.4

The half-bridge inverter circuit shown in Fig. 4.10 has a load resistance of 10Ω and an inductance of $100\,mH$. The centre-tapped battery has $20\,V$ per section. The inverter frequency is $100\,Hz$. Determine the value of I_0, and $i(t)$, (a) $3\,ms$ after Q_1 switches, and (b) $4\,ms$ after Q_2 switches.

Solution

$T = 1/f = 1/100 = 10\,ms$

$T/2 = 5\,ms$

$L/R = 0.1/10 = 10\,ms$

From equation (4.6)

$$\begin{aligned}
I_0 &= (V_b/R)(1 - \exp(-RT/2L)) / (1 + \exp(-RT/2L)) \\
&= (20/10)(1 - \exp(-5/10)) / (1 + \exp(-5/10)) \\
&= 0.49\,A
\end{aligned}$$

From equation (4.5)

$$i(t) = (V_b/R)(1 - \exp(-Rt/L)) - I_0 \exp(-Rt/L)$$

$$= (20/10)(1 - \exp(-3/10)) - I_0 \exp(-3/10)$$

$$= 0.155\,\text{A}$$

From equation (4.7)

$$i(t) = I_0 \exp(-Rt/L) - (V_b/R)(1 - \exp(-RT/2L))$$

$$= 0.49 \exp(-4/10) - (20/10)(1 - \exp(-4/10))$$

$$= -0.331\,\text{A}$$

4.5 FULL-WAVE BRIDGE INVERTER

The circuit arrangement for the full-bridge inverter is shown in Fig. 4.14. If the load is purely resistive, the feedback diodes are not required. Q_1 and Q_3 are turned on for the positive half-cycle, Q_2 and Q_4 for the negative half-cycle. The load voltage value is twice that of the half-bridge, i.e. for ideal switches the full battery voltage appears across the load. The rms load current is twice that of the half-bridge, and the load power is increased by a factor of 4.

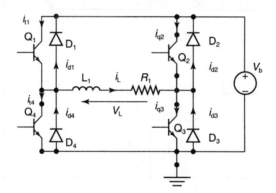

Figure 4.14

For analysis of the full-bridge, with an R–L load, equations (4.5)–(4.7) are used with the full battery voltage instead of the centre-tapped value. Figure 4.15 shows the voltage and current waveforms in the circuit.

4.6 AUXILIARY IMPULSE COMMUTATED INVERTER

When thyristors are used as the power electronic switches, a separate turn-off circuit is required in the situation where there is no natural supply voltage zero, i.e. where the supply is a steady d.c. Examples of forced commutation in inverter circuits follow.

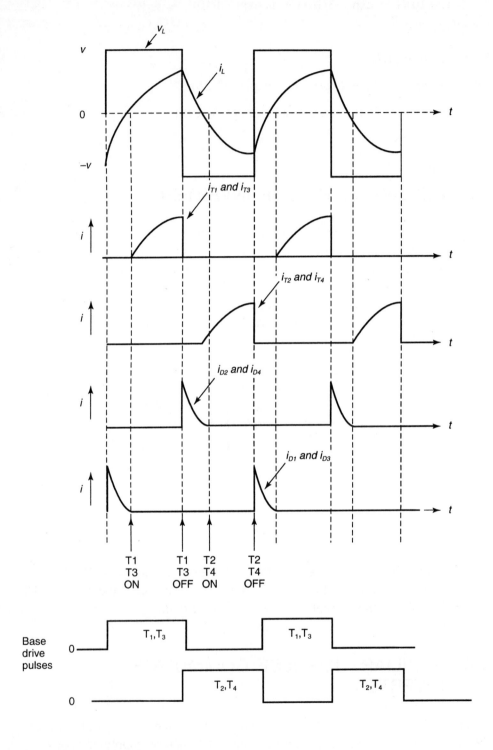

Figure 4.15   Inverter waveforms – inductive load

In Fig. 4.16, T_1 and T_2 are the main power controlling thyristors, and T_{1A} and T_{2A} are the auxiliary commutating thyristors. D_1 and D_2 are the feedback diodes for inductive loads.

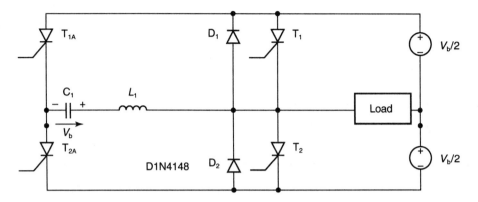

Figure 4.16

T_{2A} is switched on to pre-charge the commutating capacitor. When the charging current ceases, the capacitor is charged to the full battery voltage V_b, with the polarity shown. The thyristor turns off naturally at current zero, when the capacitor is fully charged.

T_1 is turned on to start the load cycle, and conduction occurs left to right through the load with the left-hand terminal positive. At the end of the half-cycle period, T_{1A} is turned on, T_1 is reverse-biased by the capacitor and prepares to turn off. The capacitor meanwhile recharges to opposite polarity, through the load first, and then through D_1 when T_1 turns off. When the capacitor current reaches zero, T_{1A} turns off leaving the left-hand plate of the capacitor positive.

T_2 is now switched on and conduction occurs right to left through the load with the right-hand terminal positive. At the end of the negative half-cycle period, T_{2A} is switched on, T_2 is reverse-biased and prepares to turn off, and the capacitor recharges to the original polarity through the load, and then through D_2 when T_2 turns off. The cycle is then repeated.

Sequences of gate pulses and the resulting load voltage waveform are shown in Fig. 4.17. A single firing pulse is shown, but if the load is inductive a train of pulses would be used. T_1 and T_2 must never be on together because this will short-circuit the supply. If T_2 is fired after T_1 is off, but before T_{1A} is off, the capacitor charge is topped up.

Control of inverter frequency is by control of the periodic time of the gate firing pulses. Control of the mean voltage can be achieved by variation of the delay between gating pulses of main and auxiliary thyristors. A block schematic diagram of the gate pulse control is shown in Fig. 4.18.

A typical commutation interval is shown in three stages in Figs 4.19–4.21.

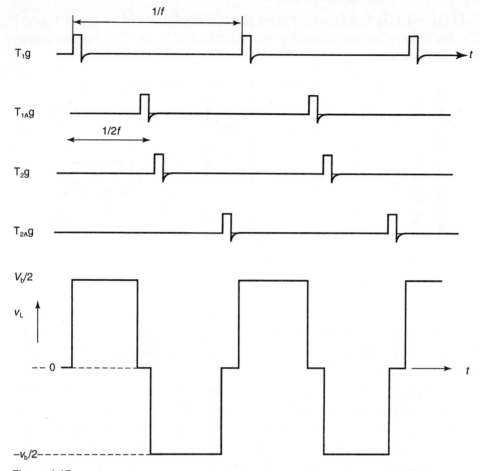

Figure 4.17

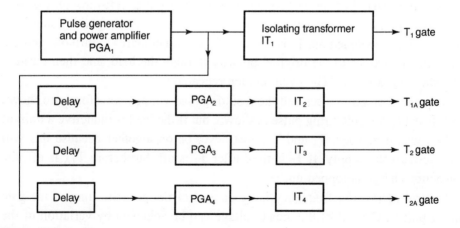

Figure 4.18

The assumption is an inductive load causing sensibly constant current over the commutation interval:

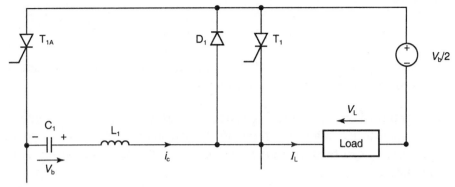

Figure 4.19

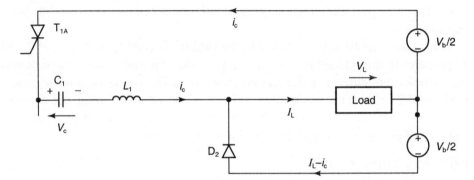

Figure 4.20

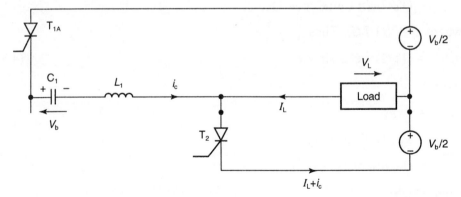

Figure 4.21

1. T_1 is on and T_{1A} is turned on to begin the commutation of T_1 (see Fig. 4.19). The capacitor discharges through an oscillatory circuit, first of all through the load, and then through D_1. When T_{1A} is turned on the capacitor current, i_c, rises and the current in thyristor T_1 falls until it reaches zero, at which point T_1 turns off. Capacitor current continues to rise, and the excess

of i_c over I_L flows through diode D_1. The discharge current passes through maximum, and begins to fall as the capacitor begins to charge in the reverse direction.

2. i_c falls below I_L, the load inductance acts to keep I_L constant, and in so doing it forward biases D_2 and load current flows in this diode (see Fig. 4.20). When the capacitor is charged in the reverse direction, i_c ceases and T_{1A} turns off, leaving the capacitor charged to rather less than V_b.

3. If T_2 is turned on just before T_{1A} goes off, a further charge will flow into C to recharge it to V_b. When T_{1A} turns off, the capacitor is charged to a polarity such that T_2 can be turned off by firing T_{2A} in the negative half-cycle (see Fig. 4.21).

Figure 4.22 shows the waveforms during a commutation interval. The resonant frequency half-periodic time is very small compared to the inverter half-periodic time.

At t_0, a commutation pulse is fed to the gate of T_{1A}. At t_1, $i_c = I$, $i_{T1} = 0$ and T_1 prepares to turn off. At t_2, $i_c = 0$ for the second time and T_1 must have turned off by this time; current is transferred from D_1 to D_2. At t_3, T_2 is turned on.

Figure 4.23 shows the position, as a circuit in the s domain, when the charged capacitor is switched across T_1. The resistance of the coil has been neglected and this of course will damp the oscillation. In this circuit:

$$i(s) \quad = V_b/2s \ (sL + 1/sC)$$

$$= V_b/2L(s^2 + 1/LC)$$

$$= (V_b/2\omega L) \ (\omega/(s^2 + \omega^2))$$

where $\omega = \sqrt{(1/LC)}$. Thus

$$i(t) \quad = (V_b/2)\sqrt{C/L} \ \sin \omega t \qquad (4.8)$$

Also

$$v_c \quad = i(s)/sC$$

$$= V_b \ /2s(s^2 \ LC + 1)$$

$$= (V_b \ /2)\omega^2/s(s^2 + \omega^2)$$

from which

$$v_c \ (t) = (V_b/2)(1 - \cos \omega t) \qquad (4.9)$$

Example 4.6

A single-phase auxiliary-impulse commutated inverter is driven from a 60 V centre-tapped d.c. supply. The load is inductive and over the commutation

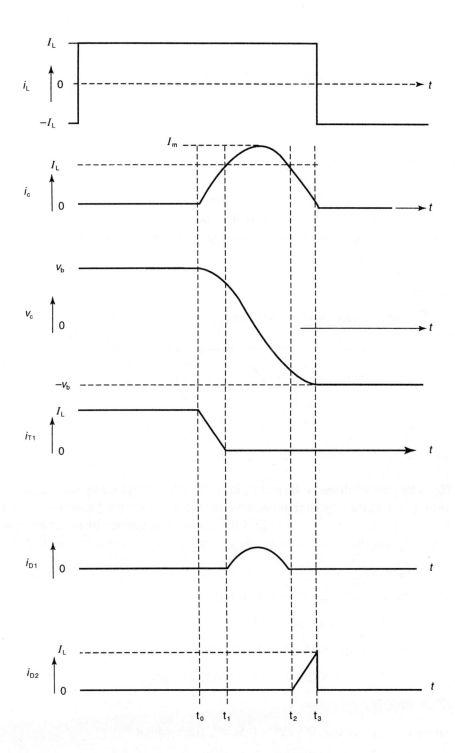

Figure 4.22

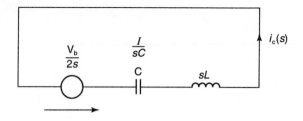

Figure 4.23

interval the load current can be considered constant at 30 A. The thyristors turn-off time is $40\,\mu s$.

Derive suitable values for the commutating components L and C. Assume that the ratio of maximum capacitor current to constant load current is 1.5.

Also determine the time taken for the capacitor current to reach the load and maximum currents.

Solution

The commutating capacitor current waveform is shown in Fig. 4.24.

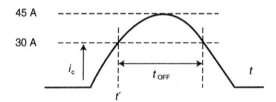

Figure 4.24

Thyristor current drops to zero at t', load current is provided by the capacitor. At time $t' + t_{off}$, the thyristor must have turned off since the capacitor current falls below I_L. The problem is to find the resonant frequency of the commutation circuit, and then the values of the commutation components L and C:

$i_c \quad = I_m \sin \omega t$

At $t = t'$, $30 = 45 \sin \omega_0 t$. Therefore $0.667 = \sin \omega_0 t'$.

$\omega_0 t' \quad = \sin^{-1} 0.667 = 41.84° = 0.73 \text{ rad}$

$\omega_0 t_{off} = 180 - 2\omega_0 t' = 96.32° = 1.682 \text{ rad}$

$\omega_0 \quad = 1.682/t_{off} = 1.682/40 \times 10^{-6} = 42\,050 \text{ rad/s}$

$\therefore f_0 \quad = 6692 \text{ Hz}$

Equating energy storage, $0.5\,LI^2 = 0.5CV^2$, from which

$L \quad = CV^2/I^2 = C\,(60)^2/(45)^2 = 1.8\,C$

Now

$\omega_0{}^2$ $= 1/LC = 1/1.8\ C^2$

C $= 1/\sqrt{1.8 \times 42\ 050} = 17.7\,\mu F$

L $= 1.8 \times 17.7 \times 10^{-6} = 31.9\,\mu H$

Time taken for current to reach 30 A is given by equation (4.8) as

$i(t)$ $= (V_b\,/2)\sqrt{C/L}\ \sin \omega t$

30     $= 45 \sin (42\ 050)t$

$\therefore t$ $= \arcsin (30/45) / 42\ 050 = 17.4\,\mu s$

Time taken for current to reach 45 A is given by equation (4.8) as

45     $= 45 \sin (42\ 050)t$

$\therefore t$ $= \arcsin (45/45)/42\ 050 = 37.4\,\mu s$

4.7   HALF-CONTROLLED BRIDGE WITH RESONANT LOAD (see Fig. 4.25)

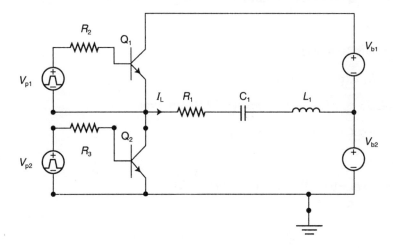

Figure 4.25

The load current response to the square wave voltage excitation of the inverter circuit with an *R–L–C* load is a sinusoid within an exponential envelope, i.e.

$i(t) = (V/\omega L) \exp(-at) \sin \omega t$

where $a = R/2L$ and $\omega^2 = (1/LC) - a^2$.

If the resonance is to continue for most of the half-cycle in order to reduce the current harmonic content, the value of $1/a = \tau$, the time constant, must be a significant part of the half-cycle period.

Example 4.7 _____

A half-controlled Mosfet inverter has a centre-tapped 40 V battery. The load is a series connected R–L–C combination with $R = 10\Omega$, $L = 25.4\,\text{mH}$ and $C = 100\,\mu\text{F}$. The inverter frequency is 100 Hz. Determine the current waveform at switch-on, and the current at $t = 2.5\,\text{ms}$ and $t = 5\,\text{ms}$.

Solution

$i(t) = (V/\omega L)\exp(-at)\sin\omega t$

where

$a = R/2L = 10/(2 \times 25.4 \times 10^{-3}) = 197$

$\tau = 1/a = 1/197 = 5.1\,\text{ms}$

$LC = 25.4 \times 10^{-3} \times 100 \times 10^{-6} = 2.54 \times 10^{-6}$

$1/LC = 39.4 \times 10^{4}\,\text{rad/s}$

$\omega^2 = (1/LC) - a^2 = (39.4 - 3.88)10^{4}$

$\omega = 596\,\text{rad/s}$

$f = \omega/2\pi = 596/2\pi = 95\,\text{Hz}$

$i(t) = (V/\omega L)\exp(-at)\sin\omega t$

$\quad = (20/596 \times 25.4 \times 10^{-3})\exp(-197t)\sin 596t$

$i(t) = 1.32\exp(-197t)\sin 596t\ (\text{A})$

At $t = 2.5\,\text{ms}$,

$i(t) = 1.32\exp(-0.493)\sin 1.43 = 0.803\ (\text{A})$

At $t = 5\,\text{ms}$,

$i(t) = 1.32\exp(-0.985)\sin 2.98 = 0.08\ (\text{A})$

Once the transients have disappeared, the current will be more or less sinusoidal with a maximum value of about 2 A. The typical load current waveform is shown in Fig. 4.26.

4.8 THREE-PHASE BRIDGE INVERTERS

A full-bridge three-phase BJT inverter is shown in Fig. 4.27. The load is a balanced star connection of pure resistors. If the load was inductive, feedback

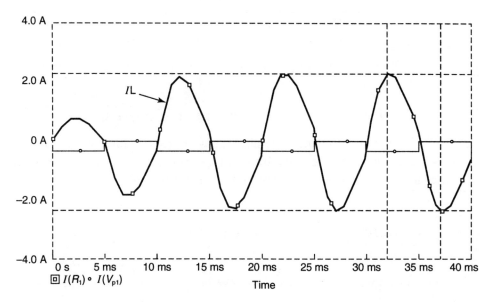

Figure 4.26

diodes would be connected across each switch; these could be built into the switch module. In this type of inverter, two or three switches could be conducting simultaneously.

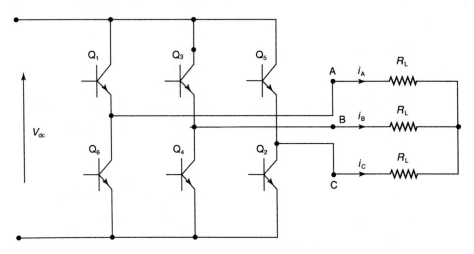

Figure 4.27

Two switches conducting

The conducting sequence is as follows (each device conducts for a 120° period):

Device	Q_1	Q_2	Q_3	Q_4	Q_5	Q_6
Device	Q_6	Q_1	Q_2	Q_3	Q_4	Q_5
θ	0–60°	60°–120°	120°–180°	180°–240°	240°–300°	300°–360°

Two conducting devices – line voltage and current

Values of line voltage and line (also phase) current for 60° intervals are shown in Fig. 4.28, from which the waveforms (Fig. 4.29) are constructed. The other two lines have identical values, displaced mutually by 120°.

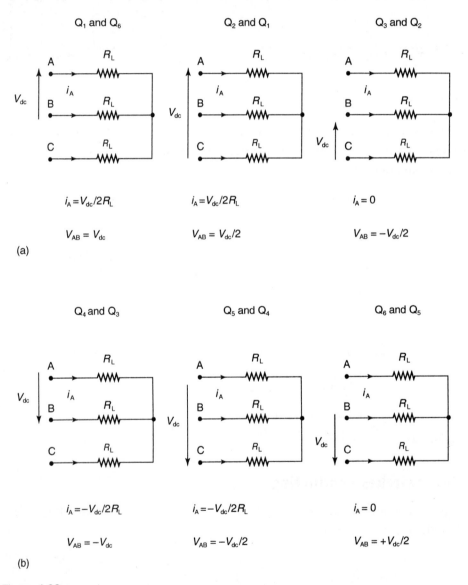

(a)

(b)

Figure 4.28

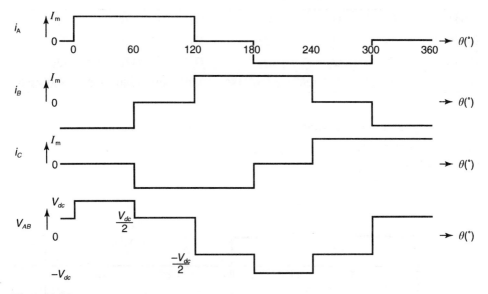

Figure 4.29

The voltage and current waveforms (Fig. 4.29) are obtained from

$$I_m = V_{dc}/2\,R_L$$

Example 4.8

A three-phase Mosfet inverter is connected between a balanced star-connected resistive load with $R_L = 10\,\Omega$ and a 250V d.c. supply. The inverter frequency is set at 33 Hz, and the switches have 120° conduction periods. Sketch and scale one of the line current waveforms, and calculate the load power.

Solution

$I_m \;=\; V_{dc}/2\,R_L = 250/2 \times 10 = 12.5\,\text{A}$

$I_{rms} \;=\; \sqrt{(1/\pi) \int_0^{2\pi/3} I_m{}^2 \delta\theta} = I_m \sqrt{(1/\pi)\,[\theta]_0^{2\pi/3}}$

$\qquad =\; I_m \sqrt{2/3} = 12.5 \sqrt{2/3} = 10.21\,\text{A}$

$P \;=\; 3\,I_{rms}{}^2\,R_L = 3\,(10.21)^2\;10 = 3.13\,\text{kW}$

Periodic time $= 1/f = 1/33 = 30\,\text{ms}$

Three switches conducting

The conducting sequence is as follows (each device conducts for a 180° period):

Device	Q_1	Q_2	Q_3	Q_4	Q_5	Q_6
Device	Q_5	Q_6	Q_1	Q_2	Q_3	Q_4
Device	Q_6	Q_1	Q_2	Q_3	Q_4	Q_5
θ	0–60°	60°–120°	120°–180°	180°–240°	240°–300°	300°–360°

The voltage and current waveforms (Fig. 4.30) are obtained from

$$I_m = 2V_{dc}/3R_L$$

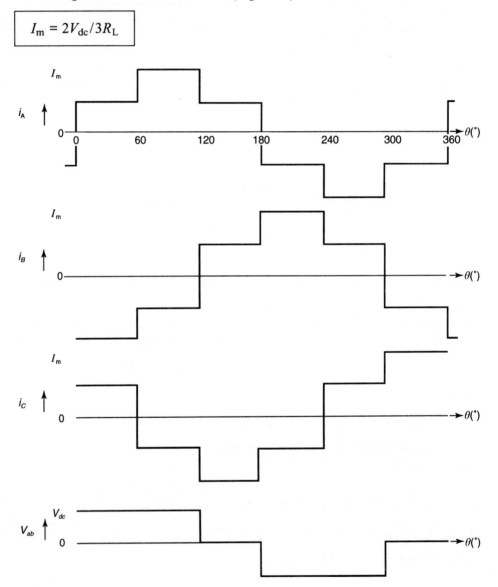

Figure 4.30

Three conducting devices – line voltage and current

Values of line voltage and line (also phase) current for 60° intervals are shown in Fig. 4.31, from which the waveforms are constructed. The other two lines have identical values, displaced mutually by 120°.

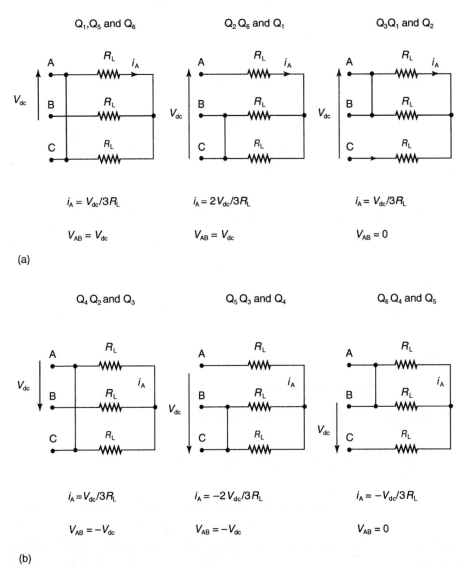

Q_1, Q_5 and Q_6

$i_A = V_{dc}/3R_L$

$V_{AB} = V_{dc}$

$Q_2 Q_6$ and Q_1

$i_A = 2V_{dc}/3R_L$

$V_{AB} = V_{dc}$

$Q_3 Q_1$ and Q_2

$i_A = V_{dc}/3R_L$

$V_{AB} = 0$

(a)

$Q_4 Q_2$ and Q_3

$i_A = V_{dc}/3R_L$

$V_{AB} = -V_{dc}$

$Q_5 Q_3$ and Q_4

$i_A = -2V_{dc}/3R_L$

$V_{AB} = -V_{dc}$

$Q_6 Q_4$ and Q_5

$i_A = -V_{dc}/3R_L$

$V_{AB} = 0$

(b)

Figure 4.31

Comparing rms line currents:

1 Two switches conducting with 120° conduction period (Fig. 4.32):

$$I_{rms} = \sqrt{1/\pi \ (V_{dc}/2R_L)^2 \ (2\pi/3)}$$

$$= V_{dc}/R_L \sqrt{(1/6)} = 0.408 \ V_{dc}/R_L \tag{4.10}$$

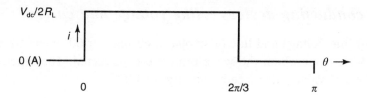

Figure 4.32

2 Three switches conducting with 180° conduction period (Fig. 4.33):

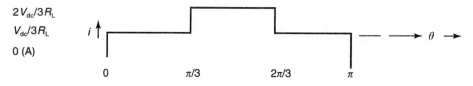

Figure 4.33

$$I_{rms} = \sqrt{1/\pi \left\{ (V_{dc}/3R_L)^2 \, (\pi/3) + (2V_{dc}/3R_L)^2 \, (\pi/3) + (V_{dc}/3R_L)^2 \, (\pi/3) \right\}}$$

$$= V_{dc}/R_L \, \sqrt{2/9} = 0.471 \, V_{dc}/R_L \tag{4.11}$$

Example 4.9 _____

Repeat Example 4.8 but using the inverter with 180° conduction period for each switch.

Solution

$$I_{rms} = 0.471 \, V_{dc}/R_L = 0.471 \times 250/10 = 11.76 \, \text{A}$$

$$P = 3I_{rms}^2 \, R_L = 3 \times 11.76^2 \times 10 = 4.15 \, \text{kW}$$

4.9   INVERTER HARMONICS

Basically the inverter voltage waveform is a symmetrical square wave. Fourier analysis shows that any non-sinusoidal waveform can be expressed as a series containing a d.c. component and a sum of sine and cosine terms, with a fundamental frequency and multiples of the fundamental frequency, called harmonics.

With a symmetrical waveform, there is no d.c. component and if the negative half-cycle is a mirror reflection of the positive half-cycle then there are no cosine terms, and only odd sine terms.

Thus the inverter waveform can be expressed in a Fourier series as

$$v = V_{1m} \sin \omega t + V_{3m} \sin 3\omega t + V_{5m} \sin 5\omega t + V_{7m} \sin 7\omega t + \ldots \ldots + V_{nm} \sin n\omega t \tag{4.12}$$

When the maximum values of the voltages are evaluated, it can be shown that equation (4.10) is equivalent to equation (4.11) below:

$$v = (4\ V_{dc}/\pi)\ (\sin\ \omega t + (1/3)\ \sin\ 3\omega t + (1/5)\ \sin\ 5\omega t + (1/7)\ \sin\ 7\omega t + \ldots + (1/n)\sin\ n\omega t) \tag{4.13}$$

This is a fundamental frequency with third, fifth, seventh up to the nth harmonic frequencies.

The overall rms value of the waveform is given by

$$V_{rms} = \sqrt{(V_{1m}^2 + V_{3m}^2 + V_{5m}^2 + V_{7m}^2 + \ldots\ldots + V_{nm}2)/2}$$

$$V_{rms} = \sqrt{(V_{1rms} + V_{3rms} + V_{5rms} + V_{7rms} + \ldots\ldots + V_{nrms})} \tag{4.14}$$

Each component of the voltage will produce a corresponding current, i.e.

$$I_{rms} = \sqrt{(I_{1rms} + I_{3rms} + I_{5rms} + I_{7rms} + \ldots\ldots + I_{nrms})} \tag{4.15}$$

The power dissipation is given by

$$P = I_{1rms}^2\ R + I_{3rms}^2\ R + I_{5rms}^2\ R + I_{7rms}^2\ R + \ldots\ldots + I_{nrms}^2\ R \tag{4.16}$$

The impedance of the load will vary with frequency, in general with R–L–C components in the load,

$$|Z_1| = \sqrt{R^2 + (\omega L - 1/\omega C)^2},\ \cos\ \phi_1 = R/|Z_1|,\ I_{1rms} = V_{1rms}\ /|Z_1|$$

Similarly

$$|Z_3| = \sqrt{R^2 + (3\omega L - 1/3\omega C)^2},\ \cos\ \phi_3 = R/|Z_3|,\ I_{3rms} = V_{3rms}\ /|Z_3|$$

$$|Z_n| = \sqrt{R^2 + (n\omega L - 1/n\omega C)^2},\ \cos\ \phi_n = R/|Z_n|,\ I_{nrms} = V_{nrms}\ /|Z_n|$$

Total harmonic distortion factor is defined as

$$\text{THD} = (\sqrt{V_{rms}^2 - V_{1rms}^2})/V_{1rms} \tag{4.17}$$

Example 4.9 _____

A single-phase full-bridge inverter is connected between a 200 V d.c. source and a series connected R–L load with $R = 20\Omega$ and $L = 50$mH. The inverter frequency is 50 Hz. Determine the values of the rms load current, the load power and the total harmonic distortion factor, up to the 9th harmonic.

Solution

$$v = (4\ V_{dc}/\pi)\ (\sin\ \omega t + (1/3)\ \sin\ 3\omega t + (1/5)\ \sin\ 5\omega t + (1/7)\ \sin\ 7\omega t + (1/n)\sin\ n\omega t)$$

$$v = (4 \times 200/\pi)(\sin \omega t + (1/3)\sin 3\omega t + (1/5)\sin 5\omega t + (1/7)\sin 7\omega t + (1/n)\sin n\omega t)$$

$$v = 255 \sin \omega t + 85 \sin 3\omega t + 51 \sin 5\omega t + 36 \sin 7\omega t + 28.3 \sin 9\omega t$$

$$\omega L = 314.2 \times 50 \times 10^{-3} = 15.71\,\Omega$$

$$Z_1 = \sqrt{20^2 + 15.71^2} = 25.4\,\Omega \qquad I_{1m} = 255/25.4 = 10\,\text{A}$$

$$Z_3 = \sqrt{20^2 + 47.1^2} = 51\,\Omega \qquad I_{3m} = 85/51 = 1.67\,\text{A}$$

$$Z_5 = \sqrt{20^2 + 78.6^2} = 81\,\Omega \qquad I_{5m} = 51/81 = 0.63\,\text{A}$$

$$Z_7 = \sqrt{20^2 + 110^2} = 112\,\Omega \qquad I_{7m} = 36/12 = 0.32\,\text{A}$$

$$Z_9 = \sqrt{20^2 + 141^2} = 143\,\Omega \qquad I_{9m} = 28.3/143 = 0.2\,\text{A}$$

$$I_{rms} = \sqrt{(10^2 + 1.67^2 + 0.63^2 + 0.32^2 + 0.2^2)/2} = 7.2\,\text{A}$$

$$P = I_{rms}^2 R = 7.2^2 \times 20 = 1037\,\text{W}$$

$$\text{THD} = (\sqrt{V_{rms}^2 - V_{1rms}^2})/V_{1rms} = \sqrt{200^2 - (255/\sqrt{2})^2}/(255/\sqrt{2}) = 0.47$$

4.10 SINUSOIDAL PULSE-WIDTH MODULATION

One of the methods used to reduce the low frequency harmonics in the inverter waveform is sinusoidal pulse-width modulation. In this method, a reference copy of the desired sinusoidal waveform, the modulating wave, is compared to a much higher frequency triangular waveform, called the carrier wave. The resulting drive signals cause multiple turn-on of the inverter switches in each half-cycle with variable pulse width to produce a quasi-sine wave of load voltage. The pulse width increases from a very narrow width at the start of each cycle to a maximum width in the middle of each cycle. Then the pulse width reduces again after maximum until its minimum width at the end of the half-cycle period.

Typically in the comparitor when the sine wave voltage exceeds the triangular wave voltage, the load voltage is $+V_{dc}$, and when the triangular wave voltage exceeds the sine wave voltage, the load voltage is $-V_{dc}$.

The magnitude of the load current can be controlled by the amplitude modulation ratio, $M_A = \text{sine } V_m/\text{triangular } V_m$. The accuracy, or closeness to a sine wave, can be controlled by the frequency modulation ratio, $M_f = f_{triangular}/f_{sine}$.

The waveforms in Figs 4.34 and 4.35 have been obtained from a PSPICE simulation, with a 200 V d.c. source, a load of $R = 20\,\Omega$ and $L = 25\,\text{mH}$, and values of $M_a = 0.8$ and $M_f = 20$.

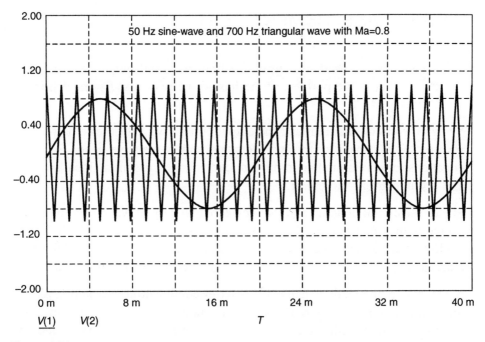

Figure 4.34

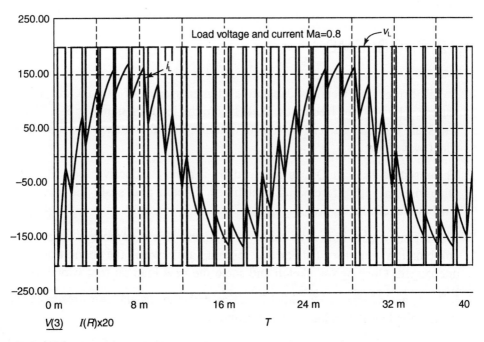

Figure 4.35

This modulating method does not eliminate the harmonics, but has the effect of shifting them to a higher frequency level where they are more easily filtered. Detailed consideration of PWM methods is beyond the scope of this book, but is dealt with by several of the books in the reference section.

4.11 SELF-ASSESSMENT TEST

1 What methods are used to control the frequency and output voltage of an inverter? What is the purpose of the base-drive resistors, R_2 and R_3, in the circuit in Fig. 4.1?

2 Sketch the circuit diagram and the output current waveform of a half-bridge inverter with capacitive bridge elements and a resistive load. State the equations for maximum and minimum load voltages, assuming ideal switches and a symmetrical waveform.

3 Sketch the circuit diagram and the typical output current waveform for a half-bridge with a centre-tapped battery voltage source and a purely inductive load. Assuming the waveform is symmetrical, what is the value of the peak load current, I_0, in terms of battery voltage, frequency and inductance?

4 In Q.3 above, the load is changed to a series combination of R and L. Sketch the shape of the load current waveform. State the expression for the load current variation with time during the on-time of each switch, and the expression required to determine I_0.

4.12 PROBLEMS

1 A half-bridge inverter has capacitive elements each of $100\,\mu F$ and a $50\,V$ battery. The inverter frequency is $1\,kHz$ and the load resistance is $5\,\Omega$. Sketch and scale the load current waveform assuming ideal switches and a symmetrical waveform.

2 The resistive load in Q.1 is replaced by a purely inductive load of $1\,mH$, otherwise the circuit remains unchanged. Sketch and scale the new load current waveform.

3 The resistive and inductive loads of Q.1 and 2 are combined in a series combination for the half-bridge circuit; otherwise the conditions are unchanged. Once again, sketch and scale the new load current waveform.

4 In Q. 1, an inductance is added in series with the resistance to resonate the load. Calculate the inductor value required.

5 A single-phase auxiliary-impulse inverter is driven from a $96\,V$ centre-tapped battery. The load current is $20\,A$, constant over the commutation interval. The maximum capacitor current is to be $30\,A$. Thyristors have a turn-off time of $50\,\mu s$. Derive suitable values for commutating capacitor and inductor.

5 AC to AC voltage regulators

Variable a.c. voltage can be obtained using Triacs and inverse-parallel connected thyristors. One method is to use phase control, where part of the sinusoidal voltage waveform is blanked out each half or full cycle. Another method is to use burst-firing control, where complete half or full cycles are blanked out. Burst-firing generates less in the way of harmonics since it switches at mains voltage zero, but it is not suitable for some loads.

Step-up or step-down of the mains input voltage is achieved using transformers, and then precise control of the rms load voltage requires an a.c. to a.c. regulator.

Triacs are used for lamp dimming, heater control and series universal motor speed control. Phase control can be used for any of these applications, but burst-firing is not suitable for mains frequency filament lamps due to lamp flicker, or for motor control due to fluctuating torque. Heater control is suited to burst-firing due to the long thermal time constant of the heating element. Other applications of Triacs are in spot welding and as solid-state contactors.

Applications requiring more than about 1000V, and 200A or so, would exceed the upper rating of Triacs, and then inverse-parallel thyristors would be used.

5.1 TRIAC PHASE CONTROLLER (Fig. 5.1)

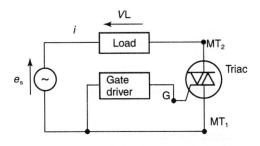

Figure 5.1

The Triac is turned on by the application of a low voltage short-duration pulse to the gate. Once on, the gate loses control and the Triac remains on until the load current falls to virtually zero, or at mains voltage zero. In the on state, there is a voltage drop of about 1 V across the Triac. As in a.c. to d.c. power control, the rms load voltage is varied using firing angle delay on mains voltage zero – the smaller the delay, the greater the rms load voltage.

5.2   RESISTIVE LOAD (see Fig. 5.2)

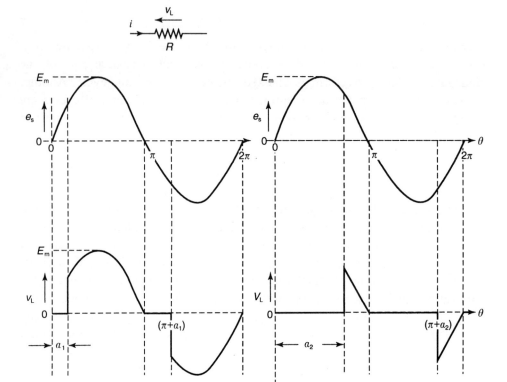

(a) Small delay angle a_1 (b) Large delay angle a_2

Figure 5.2

$$V_{rms} = \sqrt{(1/\pi) \int_a^\pi (E_m \sin \theta)^2 \, \delta\theta}$$

This will give the same value as proved for the full-wave bridge on page 37, i.e.

$$V_{rms} = (E_s) \sqrt{(1 - a/\pi + (\sin 2a)/2\pi)}$$

$$I_{rms} = V_{rms} / R$$

Example 5.1 —————————————————————

A Triac controller as shown in Fig. 5.1 provides variable power to a 200Ω resistive load from a 240V, 50Hz supply. Determine the values of rms load current, power and converter power factor for firing angle delays of (a) $\alpha = 30°$, (b) $\alpha = 150°$.

Solution

(a) I_{rms} $= V_{rms}/R = (E_s/R) \sqrt{(1 - \alpha/\pi + (\sin 2\alpha)/2\pi)}$

   $= (240/200) \sqrt{(1 - (30°/180°) + (\sin 60°)/2\pi)} = 1.18\,A$

   P $= I_{rms}^2 R = (1.18)^2 \times 200 = 278\,W$

   $\cos \phi = P/E_s I_{rms} = 278/240 \times 1.18 = 0.98$

(b) I_{rms} $= V_{rms}/R = (E_s/R) \sqrt{(1 - \alpha/\pi + (\sin 2\alpha)/2\pi)}$

   $= (240/200) \sqrt{(1 - (150°/180°) + (\sin 300°)/2\pi)} = 0.204\,A$

   P $= I_{rms}^2 R = (0.204)^2 \times 200 = 8.3\,W$

   $\cos \phi = P/E_s I_{rms} = 8.3 / 240 \times 0.204 = 0.17$

5.3 INDUCTIVE LOAD

The load is a finite inductance, $L(H)$, with a resistance $R(\Omega)$ (Fig. 5.3):

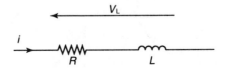

Figure 5.3

$v_L = iR + L\,\delta i / \delta t = E_m \sin (\omega t + \alpha)$

The solution to this equation for the circuit current is given as

$i = (E_m/|Z|) \{ \sin (\omega t + \alpha - \phi) + \sin (\phi - \alpha) \exp (-Rt/L)\}$ (5.2)

where

$|Z| = \sqrt{R^2 + (\omega L)^2}$

$\phi = \tan^{-1} (\omega L/R)$

The shape of the current waveform depends on the relative values of α and ϕ:

(a) $a = \phi$

Substitution of $a = \phi$ in equation (5.2) gives $i = (E_m/|Z|) \sin \omega t$. This is a pure sine wave lagging the load voltage by phase angle ϕ. The waveform is shown in Fig. 5.4.

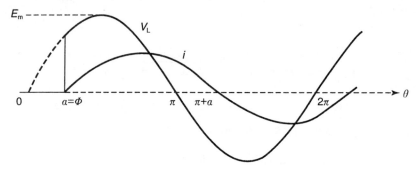

Figure 5.4

(b) $a > \phi$.

Load current starts to rise at a when the Triac switches on. Due to load inductance, the current will continue to flow after mains voltage zero. The Triac will turn off at load current zero. Difficulty of this analysis is finding the angle β for current zero, but an iterative method such as Newton–Raphson can be used for this. The waveforms are shown in Fig. 5.5.

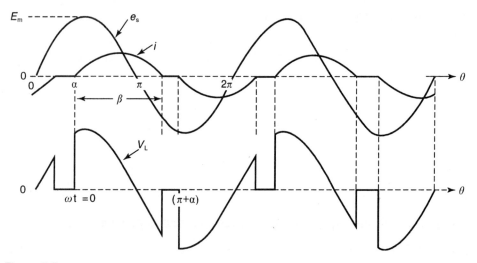

Figure 5.5

(c) $a < \phi$.

The gate pulse will arrive before current zero, and the Triac will not turn on. The solution is to use a train of pulses, and then the Triac will turn on at the first pulse after current zero.

Returning to the operation when $a > \phi$, the Triac will turn off at load current zero. The difficulty of this analysis is finding the angle β for current zero. Equation (5.2) gives the current as

$$i = (E_m/|Z|) \{ \sin (\omega t + \alpha - \phi) + \sin (\phi - \alpha) \exp(-Rt/L) \}$$

Current zero will occur when

$$\sin (\omega t + \alpha - \phi) = - \sin (\phi - \alpha) \exp(-Rt/L)$$

and this requires either an iterative approach or solution by computer simulation.

Analysis of regulator

Average values of load voltage and current are both zero, since the waveforms are symmetrical with equal areas above and below the zero voltage axis.
The root-mean-square value of the load voltage is given by

$$V_{rms} = \sqrt{(1/\pi) \int_{\alpha}^{\alpha+\beta} (E_m \sin \theta)^2 \, \delta\theta}$$

$$= E_m \sqrt{(1/\pi) \int_{\alpha}^{\alpha+\beta} 0.5(1-\cos2\theta) \, \delta\theta}$$

$$= (E_m/\sqrt{2}) \sqrt{(1/\pi) [\theta - (\sin2\theta)/2]_{\alpha}^{\alpha+\beta}}$$

$$= E_s \sqrt{(1/\pi)((\alpha + \beta) - \alpha - (\sin2(\alpha + \beta)/2) + (\sin 2\alpha)/2)}$$

$$\boxed{V_{rms} = E_s \sqrt{(\beta/\pi) - (\sin 2(\alpha+\beta)/2\pi) + (\sin 2\alpha)/2\pi}} \tag{5.4}$$

Example 5.2

The Triac regulator of Fig. 5.1 has a supply of 240V, 50Hz, and an inductive load of 15.9mH inductance and 10Ω resistance. The firing angle delay $\alpha = 60°$. Determine the value of the circuit current 4ms after firing, the angle β at which current zero occurs and the rms value of the load voltage.

Solution

Load inductive reactance

$$X_L = \omega L = 2\pi f L = 6.284 \times 50 \times 15.9 \times 10^{-3} = 5\Omega$$

$$|Z| = \sqrt{R^2 + (\omega L)^2} = \sqrt{10^2 + 5^2} = 11.18\Omega$$

$$\phi = \tan^{-1} (\omega L/R) = \tan^{-1} (5/10) = 26.56°$$

$\omega t = 314.2 \times 4 \times 10^{-3} = 1.256 \text{rad} = 71.9°$

$i = (E_m/|Z|) \{ \sin(\omega t + a - \phi) + \sin(\phi - a) \exp(-Rt/L) \}$

$= (\sqrt{2} \times 240/11.18) \{ \sin(71.9 + 60 - 26.6) + \sin(26.6 - 60) \exp(-10 \times 4/15.9) \}$

$= 30.35(0.965 - 0.044)$

$= 28 \text{A}$

If the load was resistive, the current would flow for a conduction angle $(\pi - a)$. Due to the inductance, the current hangs on longer. If the current was sinusoidal, it would go to zero at $(\pi - a + \phi)$. As can be seen from equation (5.2) the current is not a pure sine wave. However, as a first approximation, assume that it is, i.e.

$\beta = \omega t = (\pi - a + \phi) = 180° - 60° + 26.56° = 146.56°$

$t = 146.56 \times 2\pi / 360 \times \omega = 8.141 \text{ms}$

$i = (E_m/|Z|) \{ \sin(\omega t + a - \phi) + \sin(\phi - a) \exp(-Rt/L) \}$

$= (\sqrt{2} \times 240/11.18) \{ \sin(146.56 + 60 - 26.6) + \sin(26.6 - 60)$
$\exp(-10 \times 8.141/15.9) \}$

$= 30.35(\sin 180° - 0.55 \exp(-5.12))$

$= 30.35 (-0.0033) = -100 \text{mA}$

Hence the angle is very slightly less than that estimated, but this looks to be a good approximation:

$V_{rms} = E_s \sqrt{(\beta/\pi) - (\sin 2(a+\beta)/2\pi) + (\sin 2a)/2\pi}$

$= 240 \sqrt{(146.5/180) - (\sin 413°)/2\pi + (\sin 120°)/2\pi}$

$= 240 \sqrt{(0.814) - (0.127) + (0.138)}$

$= 218 \text{V}$

An *Excel* plot of the current variation of equation (5.2) for the cycle is shown in Fig. 5.6. This shows current zero at about 205° when in fact it should be $(60° + 146.6°) = 206.6°$. This error is due to the course stepping itervals of 15° chosen for the plot. The overshoot of the axis at current zero would not happen in practice since the Triac would turn off, but here it is an artifice to estimate the conduction angle. The current waveform is repeated in the negative direction starting at $(\pi + a) = 240°$.

Note the close agreement of the waveform with the calculation, $4 \text{ms} = 72°$ after firing. The calculation gives 28A, and the waveform also gives about 28A.

The plot of the current waveform for the positive half-cycle only is shown in

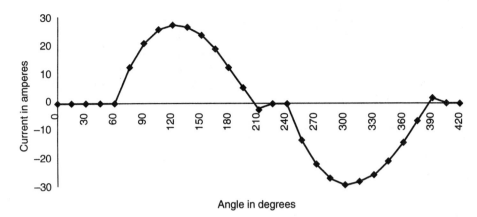

Figure 5.6

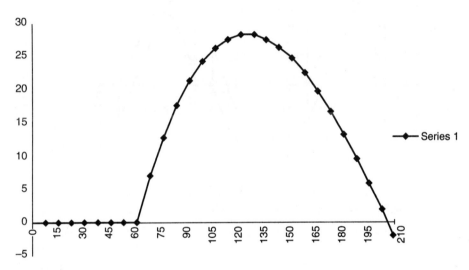

Figure 5.7

Fig. 5.7 with a stepping interval of 7.5°. This shows current zero at 206°, closer to the calculated value.

A numerical integration method, using an *Excel* spreadsheet, gives the rms value of the current as I_{rms} = 18.4 A.

Load power = $I_{rms}^2 R = (18.4)^2 \times 10 = 3.39$ kW

A PSPICE simulation of Example 5.2 is shown in Figs 5.8–5.10, Fig. 5.8 gives the circuit diagram, Fig. 5.9 gives the instantaneous load voltage and resistor voltage, and Fig. 5.10 the instantaneous load voltage and the rms resistor voltage. The voltage pulse generator switches on the Triac at α = 60° and at α = 150° in each supply cycle.

The simulation shows that current zero occurs at about 206° and the resistor rms voltage is 183 V, i.e. a current of 18.3 A, confirming calculations.

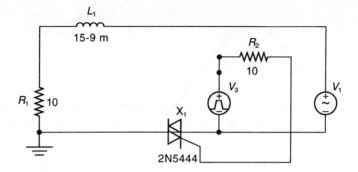

Figure 5.8

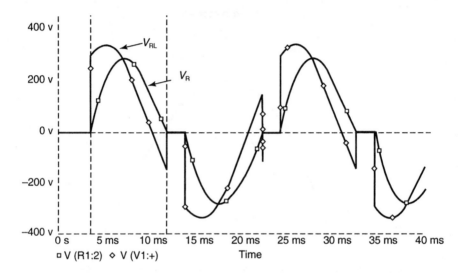

Figure 5.9

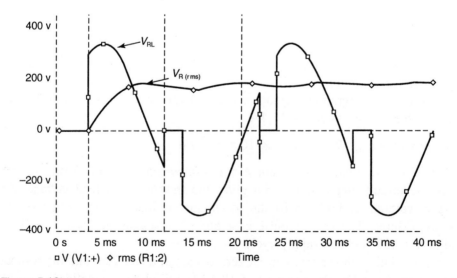

Figure 5.10

5.4   SERIES UNIVERSAL MOTOR LOAD

The series universal motor has laminated yoke and armature and can be used on
d.c. and a.c. Both pole flux and armature current reverse each half-cycle,
maintaining a unidirectional torque.

A thyristor can be used to act as a controlled half-wave voltage regulator to
provide speed and torque control of the motor. For smoother control, a Triac
will provide controlled full-wave voltage regulation.

Applications of series universal motors are in portable drills, washing
machines, vacuum cleaners etc. The circuit diagram of a half-wave regulator
is shown in Fig. 5.11.

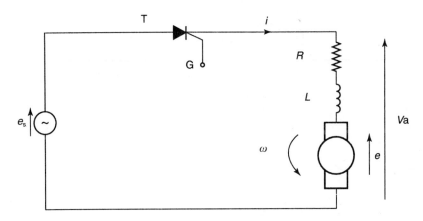

Figure 5.11

Remember that in the SEDC motor, motor flux is held constant and armature-
generated voltage depends on speed. In the series, motor flux depends on
armature current, and therefore generated voltage depends on both speed and
armature current.

The magnetic circuit will be unsaturated with light loads on the shaft, in
which case at constant speed generated voltage will be proportional to armature
current. For heavy loads, the increased field ampere-turns will cause magnetic
saturation, and then at constant speed the generated voltage will remain sensibly
constant during conduction periods.

Motor performance is governed by the following equations:

$$v_a = E_m \sin (\omega t + a) = i\,R + L\delta i/\delta t + e \quad \text{(V)} \tag{5.5}$$

$$e = k_v\,\phi\,\omega \quad\quad\quad\quad\quad\quad\quad \text{(V)} \tag{5.6}$$

$$\phi = k\,i \quad\quad\quad\quad\quad\quad\quad\quad \text{(Wb)} \tag{5.7}$$

$$t = k_t\,\phi\,i = k_t\,k\,i^2 \quad\quad\quad\quad \text{(Nm)} \tag{5.8}$$

The waveforms with unsaturated and saturated armatures are shown in Figs
5.12 and 5.13, respectively. For unsaturated operation, ϕ is proportional to

current, and assuming constant speed between firing periods, e is proportional to current.

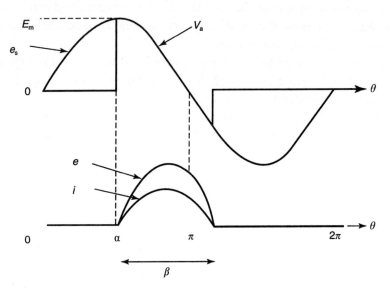

Figure 5.12

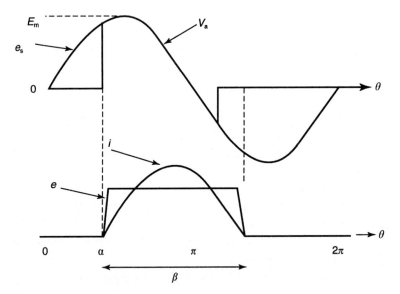

Figure 5.13

For saturated operations, ϕ is approximately constant, and if speed remains constant between firing periods, e will remain constant during the conduction period.

Example 5.3

A portable electric drill uses a series universal motor and a single thyristor. The main supply is 240V at 50Hz. Combined field and armature resistance is 16Ω and the armature voltage constant is 0.85 V/rad/s for saturated operation. Determine the average motor current and torque when the speed is 700 rev/min, the firing angle delay is 30° and the conduction angle is 180°. State any assumptions.

Solution

$$V_{av} = (1/2\pi) \int_{30}^{(30+180)} E_m \sin\theta \, \delta\theta = (E_m/2\pi)[-\cos\theta]_{30°}^{210°}$$

$$= (240/\sqrt{2}\,\pi)(-\cos 210° + \cos 30°)$$

$$= 54(0.866 + 0.866) = 93.6V$$

$$E = k_v (2\pi N/60) = 0.85 (2\pi \times 700/60) = 62.33\,V$$

$$E_{av} = E(\beta/2\pi) = 62.33(180/360) = 31.2V$$

$$I_{av} = (V_{av} - E_{av})/R = (93.6 - 31.2)/16 = 3.9\,A$$

$$T = E_{av} I_{av}/\omega = 31.2 \times 3.9 / (2\pi \times 700/60) = 1.66Nm$$

The assumptions are that the reactive voltage drop is negligible and the armature is saturated.

5.5 TRIAC BURST-FIRING CONTROLLER

Burst-firing is also known as 'on–off control', 'integral-cycle control' and 'cycle syncopation'. The method allows a number of complete supply voltage cycles (or half-cycles) through to the load and then blanks out other cycles. The power switch is turned on at mains voltage zero. The gate pulse burst firing circuit in its simplest form consists of a zero voltage detector, a pulse width modulator and logic circuitry. The basic Triac burst-firing controller circuit is shown in Fig. 5.14.

Analysis of the basic controller performance is simple, since switching is at voltage zero and rms values are readily obtained, without the need for numerical methods. The typical burst firing voltage waveform is shown in Fig. 5.15.

Root-mean-square load voltage and current:

$$V_{rms} = \sqrt{(1/m2\pi) \int_0^{n2\pi} E_m^2 \sin^2 \theta \delta\theta}$$

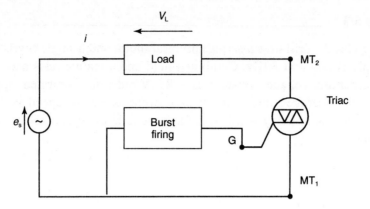

Figure 5.14

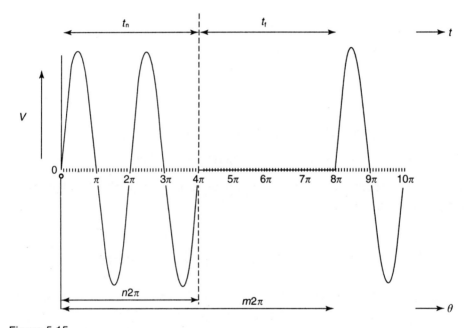

Figure 5.15

$$= \sqrt{(E_m^2/m2\pi) \int_0^{n2\pi} 0.5(1 - \cos2\theta)\delta\theta}$$

$$= \sqrt{(E_m^2/m2\pi) \int_0^{n2\pi} 0.5[\theta - (\sin2\theta)/2]}$$

$$= \sqrt{(E_m^2/m2\pi)(0.5n\ 2\pi)}$$

$$= (E_m/\sqrt{2})\sqrt{(n/m)}$$

$$\boxed{V_{rms} = E_s \sqrt{t_n/(t_n + t_f)}} \qquad (5.9)$$

Example 5.4

A burst-firing a.c. to a.c. power controller is connected between a 250V, 50Hz supply and a 100Ω resistive load. Determine the values of load voltage, current and power when (a) t_n = 40 ms and t_f = 120 ms, (b) t_n = 120 ms and t_f = 40 ms.
   Sketch load voltage waveforms for each case.

Solution

(a) V_{rms} $= E_s \sqrt{t_n/(t_n + t_f)} = 250 \sqrt{40/(40 + 120)} = 125V$

   I_{rms} $= V_{rms}/R = 125/100 = 1.25\,A$

   P $= I_{rms}^2 R = 1.25^2 \times 100 = 156.3\,W$

   $\cos \phi = P/E_s I_{rms} = 156.3/250 \times 1.25 = 0.5$

(b) V_{rms} $= E_s \sqrt{t_n/(t_n + t_f)} = 250 \sqrt{120/(40 + 120)} = 216.5\,V$

   I_{rms} $= V_{rms}/R = 216.5/100 = 2.165\,A$

   P $= I_{rms}^2 R = 2.165^2 \times 100 = 469\,W$

   $\cos \phi = P/E_s I_{rms} = 469/250 \times 2.165 = 0.867$

Burst-firing load voltage waveforms

Two cycles on/four cycles off (see Fig. 5.16)

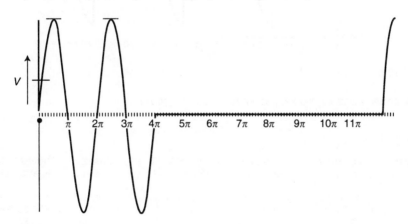

Figure 5.16

Time for one cycle is

$\theta = \omega t$

$t = \theta / \omega$

$t = 2\pi/ 2\pi f = 1/f = 1/50 = 20\,ms$

Four cycles on/two cycles off (see Fig. 5.17)

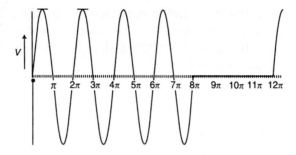

Figure 5.17

Comparison of phase control with burst-firing

A PSPICE harmonic analysis comparison has been carried out on a Triac driving a purely resistive load with (a) phase control and firing angle delay,

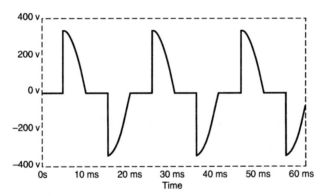

Figure 5.18 a) Waveform

```
FOURIER COMPONENTS OF TRANSIENT RESPONSE V($N_0002)


DC COMPONENT =   5.493018E-03
```

HARMONIC NO	FREQUENCY (HZ)	FOURIER COMPONENT	NORMALIZED COMPONENT	PHASE (DEG)	NORMALIZED PHASE (DEG)
1	5.000E+01	1.971E+02	1.000E+00	−3.288E+01	0.000E+00
2	1.000E+02	8.854E−03	4.491E−05	−8.578E+01	−5.290E+01
3	1.500E+02	1.076E+02	5.460E−01	8.857E+01	1.215E+02
4	2.000E+02	1.148E−02	5.822E−05	9.359E+01	1.265E+02
5	2.500E+02	3.646E+01	1.849E−01	−9.540E+01	−6.252E+01
6	3.000E+02	9.286E−03	4.710E−05	−8.874E+01	−5.586E+01
7	3.500E+02	3.592E+01	1.822E−01	8.442E+01	1.171E+02
8	4.000E+02	1.295E−02	6.570E−05	9.586E+01	1.287E+02
9	4.500E+02	2.157E+01	1.094E−01	−9.876E+01	−6.588E+01
10	5.000E+02	1.026E−02	5.203E−05	−9.936E+01	−6.648E+01

```
TOTAL HARMONIC DISTORTION =   6.143820E+01 PERCENT
```

Figure 5.18 b) Harmonics

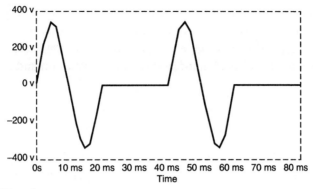

Figure 5.19 a) Waveform

FOURIER COMPONENTS OF TRANSIENT RESPONSE V($N_0001)

DC COMPONENT =   4.849595E-04

HARMONIC NO	FREQUENCY (HZ)	FOURIER COMPONENT	NORMALIZED COMPONENT	PHASE (DEG)	NORMALIZED PHASE (DEG)
1	5.000E+01	4.756E-01	1.000E+00	2.673E-01	0.000E+00
2	1.000E+02	2.216E-03	4.660E-03	-2.388E+01	-2.415E+01
3	1.500E+02	1.285E-03	2.702E-03	-6.016E+01	-6.043E+01
4	2.000E+02	5.745E-04	1.208E-03	-1.366E+02	-1.368E+02
5	2.500E+02	2.575E-04	5.413E-04	7.178E+01	7.152E+01
6	3.000E+02	1.233E-03	2.591E-03	5.146E+01	5.120E+01
7	3.500E+02	1.199E-03	2.521E-03	-2.204E+01	-2.230E+01
8	4.000E+02	7.189E-04	1.512E-03	-1.971E+01	-1.997E+01
9	4.500E+02	8.791E-04	1.848E-03	-1.321E+02	-1.324E+02
10	5.000E+02	1.181E-03	2.484E-03	8.578E+01	8.551E+01

TOTAL HARMONIC DISTORTION =   7.463852E+01 PERCENT

Figure 5.19 b)   Harmonics

$\alpha = 90°$, and (b) burst-firing control with one complete cycle on and one complete cycle off.

The analysis gives a total harmonic distortion of about 61% for phase control compared with only about 1% for burst firing. The load voltage waveforms are given in Figs 5.18 and 5.19.

Commercially produced gate drivers are available to switch on the Triac at the appropriate time, such as the 'TDA2086a' for phase control and the '443A' for burst-firing control. These gate drivers can also be built up from discrete components.

5.6   SELF-ASSESSMENT TEST

1 What are the essential differences between gate drivers for phase control and burst-firing in a.c. to a.c. voltage regulators?

2 State the equation for the rms load voltage of an a.c. to a.c. voltage regulator driving a resistive load when load power control is by (a) phase control, (b) burst-firing. Sketch typical load voltage waveforms (c) for phase control with firing angle delay $\alpha = 90°$, (d) for burst-firing with two cycles on and two cycles off.

3 State the equations for instantaneous load current of a phase-controlled a.c. to a.c. voltage regulator driving a partly inductive load. Sketch typical load voltage and current waveforms when (a) $\alpha = \phi$, (b) $\alpha > \phi$.

4 A series universal motor is driven by a thyristor half-wave regulator. Sketch armature voltage, generated voltage and motor current waveforms for (a) unsaturated armature operation, (b) saturated armature operation.

5 State the advantage and disadvantage of using burst-firing instead of phase control for a.c. to a.c. voltage regulation.

5.7   PROBLEMS

1 A phase-controlled Triac is used to control the a.c. voltage across a 125Ω resistive load. The a.c. supply is $250\,V$ at $50\,Hz$. Calculate the values of rms load voltage and current, load power and regulator power factor for firing angle delays of (a) $30°$, (b) $150°$.

2 A burst-firing Triac regulates the voltage across a 50Ω load. The a.c. input voltage is $120\,V$, $50\,Hz$. Determine the values of rms load voltage and current, load power and regulator power factor for (a) one cycle on and four cycles off, (b) four cycles on and one cycle off.

3 $250\,V$, $50\,Hz$ mains is connected to a 15Ω heating element via a Triac regulator. Calculate the load power when (a) the Triac is phase-controlled and firing angle delay $\alpha = 90°$, (b) the Triac is burst-fired with two cycles on and two cycles off. Which method is preferred?

4 Inverse-parallel connected thyristors are used to control the flow of power to a load of 80Ω resistance and $191\,mH$ inductance. The mains supply is $240\,V$ at $50\,Hz$. Using an iterative process such as Newton–Raphson, determine the conduction angle β, when firing angle delay $\alpha = 45°$.

5 Half-wave control by a single thyristor is used in a drive for a small portable drill to control the speed and torque of a series universal motor. The armature resistance is 2Ω. When the drill is operating from $240\,V$, $50\,Hz$ mains, the speed is 1400 rev/min, the firing angle delay $\alpha = 30°$, and thyristor conduction ceases at $210°$. Assuming saturated operation and an armature voltage constant $k_v = 1.2$ V/rad/s in the saturated condition, determine the average motor current and torque.

6   The DC link inverter

The d.c. link inverter is used in the situation where an a.c. supply of fixed voltage and frequency is available, but where the load requires variable voltage and variable frequency, such as speed and torque control of a squirrel cage induction motor. The a.c. source may be either three-phase or single-phase depending on the power requirements of the load. The input a.c. could use a full-wave diode rectifier system to produce fixed d.c. voltage, followed by a chopper to vary the average input voltage to the inverter. Alternatively, it could use a full-wave thyristor converter to provide variable d.c. voltage to the inverter. The circuit diagram shown in Fig. 6.1 shows a three-phase fixed voltage fixed frequency source connected to a three-phase thyristor converter. The filter smoothes out the d.c. ripple component, and the d.c. variable voltage is fed to a thyristor inverter. Today the inverter thyristors would be replaced by Mosfets in all but the highest power requirements.

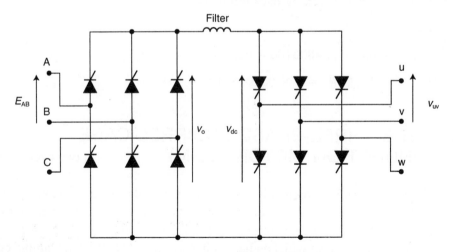

Figure 6.1

From equation (3.21) for a fully controlled converter

$$V_{o(av)} = (3E_{Lm}/\pi)\cos\alpha \tag{6.1}$$

Example 6.1 _____

A d.c. link inverter has an a.c. mains rms line input of 415 V, 50 Hz. The inverter load is a balanced star connection of 100 Ω resistors. Determine load current and power for a converter firing angle delay of (a) 50° and (b) 85°, with an inverter frequency of 100 Hz and a 120° conduction period of the inverter switches.

Solution

(a) a = 50°.

$V_{o(av)}$ = $(3 \times \sqrt{2} \times 415/\pi)$ cos 50° = 360 V

Load current is given by equation (4.11) as

I_{rms} = 0.471 V_{dc}/R_L

I_{rms} = 0.471 × 360/100 = 1.7 A

P = 3 $I_{rms}^2 R_L$ = 3 × 1.7² × 100 = 863 W

(b) a = 85°

$V_{o(av)}$ = $(3 \times \sqrt{2} \times 415/\pi)$ cos 85° = 48.8 V

Load current is given by equation (4.11) as

I_{rms} = 0.471 V_{dc}/R_L

I_{rms} = 0.471 × 48.8/100 = 0.23 A

P = 3 $I_{rms}^2 R_L$ = 3 × 0.23² × 100 = 16 W

6.1 THE THREE-PHASE SQUIRREL-CAGE INDUCTION MOTOR (SCIM)

The d.c. link inverter is frequently used to drive SCIMs in a.c. variable speed drives. Typical applications are electric road vehicles, trains etc. The stator winding of the SCIM is a conventional three-phase winding. The rotor winding has the squirrel cage form with conductors housed in slots around the laminated iron core. The rotor winding is permanently closed in on itself by means of shorting end-rings. No external connection is possible to the rotor with this construction. This is one of the simplest and cheapest of motors, without the need for brushes, slip-rings or commutator, giving a very long maintenance-free operating life. A simplified view of the machine construction is shown in Fig. 6.2.

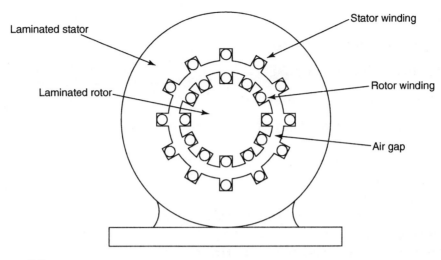

Figure 6.2

When a three-phase supply is connected to the stator winding, a field is produced, rotating around the stator at synchronous speed. The relationship between a.c. supply frequency, f(Hz), the synchronous speed of the rotating flux, n_s (rev/s) and the number of pole-pairs on the stator, p, is $f = n_s \times P$.

   For 50 Hz and a two-pole stator

$$n_s = f/p = 50/1 = 50 \text{ rev/s}$$

$$N_s = 3000 \text{ rev/min}$$

For 50 Hz and an eight-pole stator

$$n_s = f/p = 50/4 = 12.5 \text{ rev/s}$$

$$N_s = 750 \text{ rev/min}$$

The rotating field will cut the rotor conductors and induce a voltage in them. The magnitude of this voltage will depend on the relative number of rotor to stator turns. In fact the motor is acting like a transformer, inefficient because of the air gap. The rotor-induced voltage will circulate a current in the rotor conductors and this will react with the field to produce force, torque and rotation. The rotor builds up speed, and as it does so, the relative rate at which the stator field cuts the rotor conductors is reduced. This in turn reduces rotor voltage, current and torque (it also reduces the frequency of the rotor current). If the rotor reached synchronous speed, there would be no movement of stator field relative to the rotor conductors, and no induced voltage, current or torque. In order that the no-load losses can be met, the rotor attains a speed less than synchronous. When the motor shaft is loaded, the speed falls to increase induced voltage, current and torque.

Let

synchronous speed = n_s (rev/s)

actual rotor speed = n_r (rev/s)

rotor slip speed   = $n_s - n_r$ (rev/s)

The fractional slip s is defined as the ratio of slip speed to synchronous speed, i.e.

$$s = (n_s - n_r)/n_s \qquad\qquad (6.2)$$

If a 50 Hz, two-pole induction motor had a slip of $s = 0.025$, what would be the rotor speed?

From equation (6.2) $sn_s = n_s - n_r$. Hence

n_r $= n_s(1 - s) = (f/p)(1-s)$

n_r $= (50/1)(1 - 0.025) = 48.75$ rev/s

$\therefore N_r = 2925$ rev/min

Rotor induced voltage, E_r

Let E_{ro} be the rotor-induced voltage with the rotor stationary.
The rotor-induced voltage will be proportional to the rate at which the stator field cuts the rotor conductors, i.e.

$E_{ro}/E_r = n_s / (n_s - n_r)$

$$E_r = E_{ro} (n_s - n_r)/n_s = sE_{ro} \qquad\qquad (6.3)$$

Frequency of rotor current, f_r

The frequency of the rotor current will also depend on the slip since $f = n_s p$ and rotor frequency $f_r = (n_s - n_r)p$:

$$f_r = f (n_s - n_r)/n_s = sf \qquad\qquad (6.4)$$

Rotor circuit reactance, X_r

The rotor circuit impedance will have resistance R_r and reactance X_r. Let the reactance of the stationary rotor be X_{ro}. The reactance will vary with frequency, at any running speed $(f_r/f) = (X_r/X_{ro})$:

$$X_r = (f_r X_{ro}/f) = sX_{ro}$$

6.2 SCIM PHASE EQUIVALENT CIRCUIT

The stator circuit will have winding resistance R_s and reactance X_s. With no load on the rotor shaft, a current flows in a resistor R_o to meet the hysteresis and eddy current losses, and in a reactance X_o to magnetize the magnetic circuit.

The rotor circuit will have winding resistance R_r and reactance X_r. The phase equivalent circuit is similar to that of the transformer as shown in Fig. 6.3.

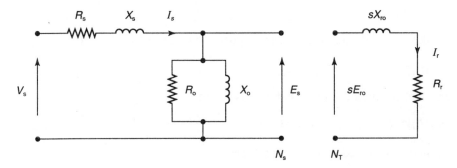

Figure 6.3

For the rotor circuit

$$sE_{ro} = (R_r + j\, sX_{ro})I_r$$
$$E_{ro} = ((R_r/s) + jX_{ro})I_r$$

This results in a modified rotor circuit as given in Fig. 6.4, where

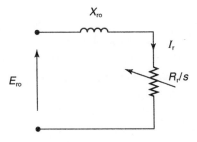

Figure 6.4

$E_s/E_{ro} = N_s/N_r = n$ (effective turns ratio)

Finally all quantities can be referred to the stator to give the circuit in Fig. 6.5. In this circuit

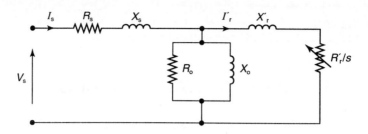

Figure 6.5

$R'_r/s = n^2 R_r/s$

$X'_r = n^2 X_r$

$I'_r = nI_r$

From the referred equivalent circuit, power transferred to the rotor is

$P_r = |I'_r|^2 R'_r/s$ (W/phase)

power lost in rotor circuit resistance is

$P_c = |I'_r|^2 R'_r$ (W/phase)

and power converted to mechanical use is

$P_m = P_r - P_c$

i.e.

$$P_m = (|I'_r|^2 R'_r/s) (1 - s)$$ (W/phase)   (6.6)

Ratio $P_r: P_c: P_m = 1:s:(1-s)$

Torque,

$$T_{ph} = P_m/\omega_r = P_m/2\pi n_r = P_m/2\pi n_s (1 - s)$$ (6.7)

Equations (6.6) and (6.7) give torque per phase as

$T_{ph} = (|I'|^2 R'_r/s \, 2\pi n_s)$

and total torque as

$$T = 3 |I'_r|^2 R'_r/s \, 2\pi n_s$$ (Nm)   (6.8)

The torque against slip (or rotor speed) characteristic can be plotted from equation (6.8). There are two variables, the referred rotor current I'_r and the fractional slip, s. The typical shape of this curve is shown in Fig. 6.6, at rated voltage and frequency, with a superimposed constant torque load to show the operating slip. T_0 is the starting torque and T_{L1} is a load torque. s_1 and n_{r1} are the operating slip and rotor speed to meet the load torque.

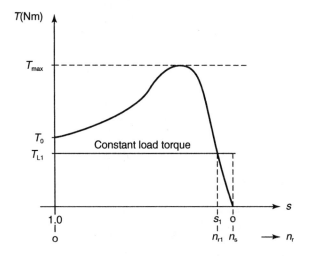

Figure 6.6

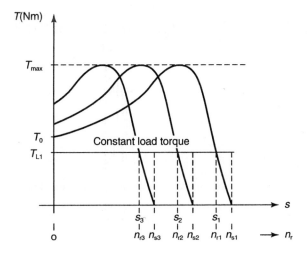

Figure 6.7

The characteristic given in Fig. 6.7 is for variable frequency operation, again with a constant torque load. However, here the supply voltage is not constant, but is varied in proportion to frequency in order to keep the motor magnetic flux constant at its rated value. Thus $V_s/f = k$. The rotor slip and speed change as the frequency reduces. The synchronous speeds at the reducing frequencies are s_1, s_2 and s_3. At low frequencies, the stator voltage drop becomes more significant and a voltage boost is necessary.

A d.c. to a.c. inverter is used to vary the motor frequency, and an a.c to d.c. converter can be used to vary the d.c. link inverter input voltage in proportion to frequency. A block diagram of a typical system is shown in Fig. 6.8.

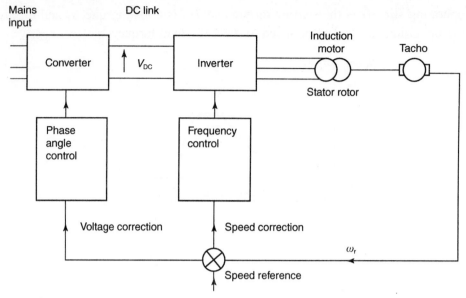

Figure 6.8   Induction motor speed control system

6.3   EFFECT OF CHANGE OF FREQUENCY ON THE EQUIVALENT CIRCUIT

Let f_1 be the rated frequency, f_2 the new frequency, V_s the rated phase voltage and V_s' the new phase voltage. Now

$V_s/f_1 \; = \; V_s'/f_2 = k$

$\therefore \; V_s' \; = \; V_s \, (f_2/f_1) = a \, V_s$

Also all motor reactances will vary in proportion to frequency:

$X_s/X_s' \; = \; f_1/f_2$

$\therefore \; X_s' \; = \; X_s \, (f_2/f_1) = a \, X_s$

The equivalent circuits at frequences f_1 and f_2 are shown in Figs 6.9 and 6.10; the core loss has been neglected.

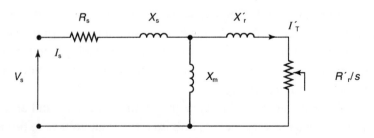

Figure 6.9   Rated frequency, f_1

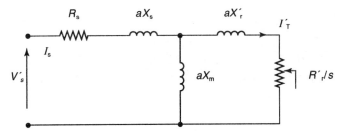

Figure 6.10   New frequency f_2, where $a = f_2/f_1$

Example 6.2

A three-phase, star-connected, four-pole, 50 Hz, 415 V, induction motor has the following phase parameters at rated frequency:

$$R_s = R_r' = 0.33\,\Omega, \; X_s = X_r' = 1.6\,\Omega, \; X_m = 20\,\Omega$$

Determine the motor currents and torque when driven by a variable voltage, variable frequency supply at 0.033 slip with constant flux (a) at rated voltage and frequency, (b) at a frequency of 25 Hz.

Assume negligible iron losses and a sinusoidal current waveform.

Solution

(a) $n_{s1} = f_1/p = 50/2 = 25$ rev/s

$N_{s1} = 25 \times 60 = 1500$ rev/min

Rotor speed, $N_r = N_{s1}(1 - s) = 1500(1 - 0.033) = 1451$ rev/min

Phase voltage $= 415/\sqrt{3} = 240$ V

Equivalent circuit at 50 Hz is shown in Fig. 6.11.

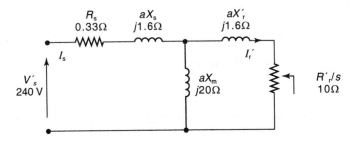

Figure 6.11

By mesh analysis:

$$
\begin{array}{ll}
\text{for stator circuit} \\
\text{for rotor circuit}
\end{array}
\begin{vmatrix} 240 \\ 0 \end{vmatrix} =
\begin{vmatrix} (0.33 + j21.6) & -j20 \\ -j20 & +(10 + j21.6) \end{vmatrix}
\begin{vmatrix} I_s \\ I_r' \end{vmatrix}
$$

Using determinates

$$I_s = \frac{\begin{vmatrix} 240 & -j20 \\ 0 & (10 + j21.6) \end{vmatrix}}{\begin{vmatrix} (0.33 + j21.6) & -j20 \\ -j20 & (10 + j21.6) \end{vmatrix}}$$

$$I_s = \frac{2400 + j5184}{3.33 + j7.13 + j216 - 466.6 + 400}$$

$$= \frac{2400 + j5184}{-63.3 + j223}$$

$$I_s = \frac{5713\lfloor 65.2°}{23.2\lfloor 105.8°} = 24.6\lfloor -40.6° \text{ (A)}$$

$$I_r' = \frac{\begin{vmatrix} (0.33 + j21.6) & 240 \\ -j20 & 0 \end{vmatrix}}{232\lfloor 105.8°} = \frac{4800\lfloor 90°}{232\lfloor 105.8°} = 20.7\lfloor -15.8° \text{ (A)}$$

$$T = 3I_r'^2 \, R_r'/s \, \omega_s = 3 \times 20.7^2 \times 10 \, / \, 2\pi \times 25 = 81.8 \text{ Nm}$$

(b)   At 25 Hz, 0.033 slip:

$$n_{s2} = f_2/p = 25/2 = 12.5 \text{ rev/s}$$

$$N_{s2} = 12.5 \times 60 = 750 \text{ rev/min}$$

Rotor speed, $N_{r2} = N_{s2}(1 - s) = 750(1 - 0.033) = 725.3$ rev/min

$$a = f_2/f_1 = 25/50 = 0.5$$

Phase voltage, $V_s' = aV_s = 0.5 \times 240 = 120 \text{V}$

$$aX_s = 0.5 \times 1.6 = 0.8\Omega = aX_r'$$

$$aX_m = 0.5 \times 20 = 10\Omega$$

Equivalent circuit at 25 Hz is shown in Fig. 6.12.

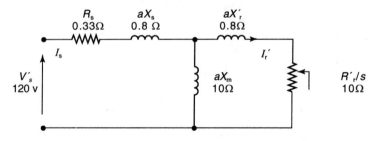

Figure 6.12

By mesh analysis:

for stator circuit $\quad \begin{vmatrix} 120 \\ 0 \end{vmatrix} = \begin{vmatrix} (0.33 + j10.8) & -j10 \\ -j10 & +(10 + j10.8) \end{vmatrix} \begin{vmatrix} I_s \\ I_r' \end{vmatrix}$

for rotor circuit

Using determinates

$I_s = \dfrac{\begin{vmatrix} 120 & -j10 \\ 0 & (10 + j10.8) \end{vmatrix}}{\begin{vmatrix} (0.33 + j10.8) & -j10 \\ -j10 & (10 + j10.8) \end{vmatrix}}$

$I_s = \dfrac{1200 + j129.6}{3.33 + j3.56 + j108 - 116.6 + 100}$

$\quad = \dfrac{1200 + j129.6}{-13.3 + j111.6}$

$I_s = \dfrac{176.6\underline{|47.2°}}{11.2\underline{|105.8°}} \quad = 15.8\underline{|-50°}$ (A)

$I_r' = \dfrac{\begin{vmatrix} (0.33 + j10.8) & 120 \\ -j10 & 0 \end{vmatrix}}{112\underline{|96.8°}} = \dfrac{1200\underline{|90°}}{112\underline{|96.8°}} = 10.7 \underline{|-7°}$ (A)

$T = 3\,I_r'^2\,R_r'/s\ \omega_s = 3 \times 10.7^2 \times 10 / 2\pi \times 12.5 = 43.7$ Nm

Example 6.3

A three-phase, 415 V, six-pole, 50 Hz, star-connected induction motor is driven from a variable voltage, variable frequency supply. The motor phase equivalent circuit parameters, at 50 Hz, referred to the stator are as follows:

$R_s = R_r' = 0.2\Omega$, $X_s = X_r' = 0.6\Omega$, $X_m = 10\Omega$

Neglecting harmonics and core losses, calculate motor currents and torque at 5% slip when the supply line voltage is (a) 415 V, 50 Hz and (b) 83 V, 10 Hz.

Solution

An alternative method of solution for the induction motor equivalent circuit to that of mesh analysis is to use series-parallel reduction. From the circuit in Fig. 6.9 the phase impedance of the motor, at rated frequency, is as given in Fig. 6.13.

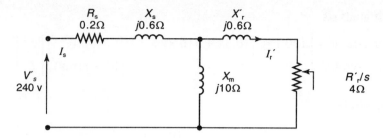

Figure 6.13

n_{s1} $=$ $f_1/p = 50/3 = 12.67$ rev/s

R_r'/s $=$ $0.2 / 0.05 = 4\Omega$

Z_m $=$ $0.2 + j0.6 + \dfrac{j10(4 + j0.6)}{4 + j10.6}$

$=$ $0.2 + j0.6 + \dfrac{(j40 - 6)(4 - j10.6)}{4^2 + 10.6^2}$

$=$ $0.2 + j0.6 + j1.25 + 3.3 - 0.19 + j0.495 = 3.31 + j2.35 \ (\Omega)$

I_s $=$ $\dfrac{240}{3.31+j2.35}$ $=$ $\dfrac{240}{4.06\lfloor 35.4°}$ $=$ $59.1\lfloor -35.4°$ (A)

I_r' $=$ $\dfrac{I_s\,(j10)}{4+j10.6}$ $=$ $\dfrac{59.1\lfloor -35.4° \times 10\lfloor 90°}{11.3\lfloor 69.3°}$ $=$ $59.1\lfloor -35.4°$ (A)

T $=$ $\dfrac{3\,I_r'^2\,R_r'/s}{2\pi n_{s1}}$ $=$ $\dfrac{3 \times 52.2^2 \times 4}{2\pi \times 16.67}$ $=$ 312 Nm

The circuit at 10 Hz with all reactances and phase voltage reduced proportional to frequency is given in Fig. 6.14. Hence

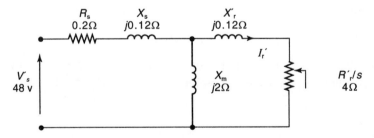

Figure 6.14

n_{s2} $=$ $f_2/p = 10/3 = 3.33$ rev/s

R_r'/s $=$ $0.2/0.05 = 4\Omega$

Z_m $=$ $0.2 + j0.12 + \dfrac{j2(4 + j0.12)}{4 + j2.12}$

$$= \ 0.2 + j0.12 + \frac{(j8 - 0.24)(4 - j2.12)}{4^2 + 2.12^2}$$

$$= \ 0.2 + j0.12 + j1.56 + 0.827 - 0.047 + j0.025 = 0.98 + j1.71 \ (\Omega)$$

$$I_s \quad = \quad \frac{48}{0.98 + j1.71} \quad = \quad \frac{48}{1.97\underline{|60.2°}} \quad = \quad 24.4\underline{|-60.2°} \ (A)$$

$$I_r' \quad = \quad \frac{I_s \ (j2)}{4 + j2.12} \quad = \quad \frac{24.4\underline{|-60.2°} \times 2\underline{|90°}}{4.53\underline{|27.9°}} \quad = \quad 10.77\underline{|-1.9°} \ (A)$$

$$T \quad = \quad \frac{3 \ I_r'^2 \ R_r'/s}{2\pi n_{s2}} \quad = \quad \frac{3 \times 10.77^2 \times 4}{2\pi \times 3.33} \quad = \quad 66.4 \ \text{Nm}$$

6.4   FURTHER STUDY TOPICS ON INDUCTION MOTOR CONTROL

Areas for further study in the field of speed and torque control of induction motors are:

1 More sinusoidal current operation using pulse-width-modulation techniques and filtering.

2 Feedback of energy from the motor to the d.c. link takes place when the motor is driven above synchronous speed by the load; the motor becomes a generator, the inverter feedback diodes are forward-biased, and current flow is reversed. If the d.c. side is a battery, such as in an electric car, the battery is charged and the motor regenerates, acting to brake the motor. When the d.c. link is fed from a controlled rectifier, it is not possible to reverse the current flow through the rectifier, but braking of the motor can be achieved by switching a resistor across the link when current flow reverses. This is called dynamic braking.

3 Vector control of the induction motor is used in modern systems to create independent control of the stator current components, causing flux and torque production. The aim is to create orthogonal axes for flux and current in a manner similar to that of the separately excited d.c. motor.

The power electronic books by Rashid (1993) Vithayathil (1995) deal in some detail with these topics.

6.5   SELF-ASSESSMENT TEST

1 Draw a circuit diagram of a d.c. link inverter system and briefly explain the operation of the circuit.

2 A three-phase, 50 Hz, induction motor has (a) 12 poles, (b) six poles, (c) two poles. Determine the synchronous speed in each case.

3 Determine the rotor speed of a three-phase, two-pole, 50 Hz, induction motor running with a slip of (a) 0.02, 9b) 0.05, (c) 3%. The motor frequency is altered to 80 Hz. What are the new rotor speeds at the above slips?

4 (a) Why is an induction motor operated in the constant flux mode during variable frequency operation?

(b) Why is a voltage boost necessary when operating the induction motor as in (a) when the frequency is low?

5 Sketch the typical torque-slip curves of a 50 Hz, two-pole, induction motor operating in the constant flux mode for 20 Hz frequency steps, from 20 to 100 Hz.

6.6   PROBLEMS

1 A three-phase, 415 V, six-pole, star-connected induction motor has the following specification: phase parameters at 50 Hz are

$R_s = R_r' = 0.2\Omega, X_s = X_r' = 0.6\Omega, X_m = 10\Omega$

Neglecting harmonics and core losses, calculate the induction motor speed, current and torque when running at 5% slip from a 166 V, 20 Hz supply.

2 A three-phase, 415 V, 50 Hz, six-pole, star-connected induction motor is driven from a variable voltage, variable frequency supply. The motor circuit phase parameters are as follows:

$R_s = R_r' = 1.0\Omega, X_s = X_r' = 2.0\Omega, X_m = 50\Omega$

When running at full load on rated voltage and frequency, the motor takes a line current of 12 A at 0.86 power factor lagging. Assuming that the motor is operating in the constant flux mode, determine the supply voltage and speed when delivering the same torque at 60 Hz.

7 Switched-mode power supplies

Switched-mode power supplies (SMPSs) are both smaller and more efficient than linear regulator power supplies. Where the output is to be d.c. and the input is a.c. mains, and isolation is required between input and output, the mains is first rectified and smoothed. The smoothed d.c. is then chopped at a high frequency (high in terms of mains frequency), i.e. typically 50 kHz. The chopped current flows in the primary of a ferrite cored transformer, much smaller than its 50 Hz counterpart. The corresponding secondary current is rectified and smoothed. Control of output voltage is obtained by adjustment of the chopper duty cycle.

The relatively high operating frequency allows values of coil inductance to be much lower than corresponding 50 Hz values, again requiring smaller size. SMPSs can produce output voltages lower or higher than the value of the smoothed d.c. link.

Standard types of SMPSs are

- buck or forward converter
- boost or flyback converter
- buck–boost converter
- cuk converter
- push–pull converter
- resonant converter
- half-bridge and full-bridge converters.

The choice of the type of SMPS to use is made on the simplicity of the drive and control circuitry on the one hand, and the power output requirements on the other.

7.1 FORWARD OR BUCK CONVERTER

Some alternative circuits for the forward converter are given in Figs 7.1–7.3.

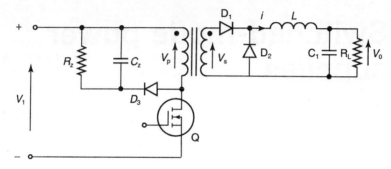

Figure 7.1

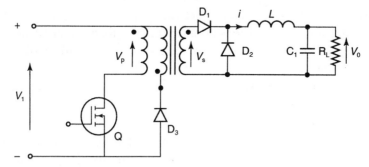

Figure 7.2

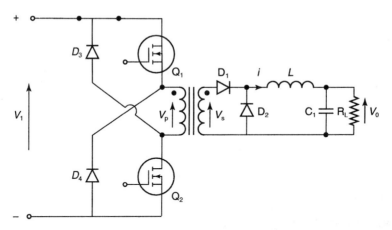

Figure 7.3

In Fig. 7.1, when T_1 is on, the secondary induced voltage forward biases diode D_1 and causes the current to ramp-up in the inductor L_1. When T_1 is off, the voltage reverses in L_1, diode D_2 is forward-biased and current flows in the loop L–C_1–D_2, replenishing the charge on the capacitor C_1 and allowing the inductor current to ramp down.

Meanwhile, the primary voltage reverses to keep the current flowing, D_3 is forward-biased and C_2 is charged up allowing the demagnetization of the transformer core. D_3 becomes reverse-biased and turns off, allowing C_2 to discharge through R_2.

An alternative to this simple core resetting circuit is to use a third demagnetizing transformer winding, as shown in Fig. 7.2. When T_1 turns off, D_3 is forward-biased, allowing current flow back to the supply to reset the core.

A third alternative is to use two switches as shown in Fig. 7.3. In this circuit two Mosfets are used, and they are switched on simultaneously. A gate-pulse isolation circuit is required. When the switches turn off, the primary induced voltage changes polarity, D_3 and D_4 are forward-biased, and the core is demagnetized as the stored energy is returned to the supply.

For the circuits given in Figs 7.1–7.3 assume that switches are ideal, that the transformer turns ratio N_p: N_s = 1, and that the smoothing capacitor C_1 is large enough to ensure a constant output voltage, V_0, between switch firing, or off periods.

During the switch-on period, $V_1 = V_p = V_s$. Hence the inductor voltage,

$L \, di/dt = V_1 - V_0$

Over the on-period, t_n, the current ramps up from I_0 to I_1:

$(V_1 - V_0)/L = di/dt = (I_1 - I_0)/t_n$

$(I_1 - I_0) = (V_1 - V_0)t_n/L$ (7.1)

During the switch off-period, $V_0 = L \, di/dt$. Over the off-period t_f, the current ramps down from I_1 to I_0. Therefore

$I_1 - I_0 = (V_0/L) \, t_f = (V_0/L) \, (T - t_n)$ (7.2)

Equating (7.1) to (7.2):

$(V_1 \, t_n/L) - (V_0 \, t_n/L) = (V_0 T/L) - (V_0 t_n/L)$

from which

$$\boxed{V_0 = DV_1}$$ (7.3)

where D is the duty cycle.

The waveform of inductor current against time is shown in Fig. 7.4.

The waveform of transformer primary voltage aginst time is shown in Fig. 7.5. If the transformer core is to be completely demagnetized in the off-period then the area above the zero voltage axis must be equal to the area below the zero voltage axis. This limits the duty cycle, D, to a value of 0.5.

If the duty cycle is greater than 0.5, the core would not be completely demagnetized at the end of the off-period, and a d.c. magnetization of the core would build up, resulting in core saturation.

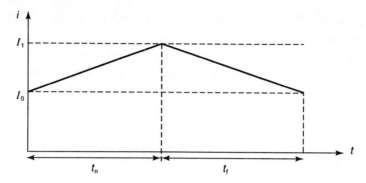

Figure 7.4

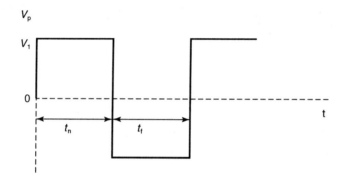

Figure 7.5

Example 7.1

An isolated forward (buck) converter has a d.c. input voltage of 30V, a 1:1 transformer and an ideal Mosfet.

The inductor value is $100\,\mu$H and and the switching frequency is 50kHz. Assume that the output capacitor is large enough to hold the load voltage constant. The converter is required to deliver 4A into a $3\,\Omega$ resistive load.

Determine the converter duty cycle, the switch on-time and the ripple current magnitude.

Solution

$V_0 = I \times R_L = 4 \times 3 = 12\,\text{V}$

Duty cycle $D = (V_0/V_1) = 12/30 = 0.4$

Periodic time $T = 1/f = 1/50 \times 10^3 = 20\,\mu\text{s}$

Switch on-time $t_n = DT = 0.4 \times 20 = 8\,\mu\text{s}$

Current ripple, $(I_1 - I_0) = (V_1 - V_0)\,t_n/L = (30 - 12)8/100 = 1.44\,\text{A}$

7.2 FLYBACK OR BOOST CONVERTER

The step-up chopper given in Chapter 2, Fig. 2.17 is the basis of the boost converter. The circuit is reproduced in Fig. 7.6 for convenience.

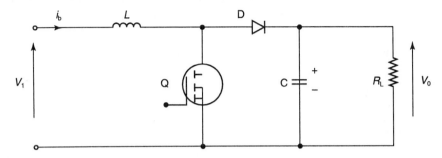

Figure 7.6

Equation (2.13) gives

$$V_0 = V_b / (1 - D)$$

and

$$I_b = I_0 / (1 - D)$$

Now consider what happens when the inductor and transistor positions are interchanged to produce the buck–boost converter.

7.3 BUCK–BOOST CONVERTER

The circuit arrangement is shown in Fig. 7.7.

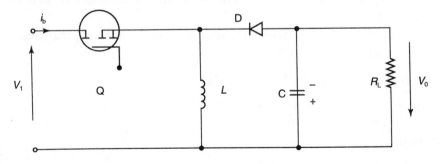

Figure 7.7

With the transistor switched on, current ramps up in the inductor, i.e.

$$V_1 = L di/dt$$
$$V_1/L = di/dt$$
$$di = (V_1/L) dt$$

At the end of the on-period,

$$I_1 - I_0 = (V_1/L)t_n \tag{7.4}$$

When the transistor is switched off the inductive voltage, V_L, changes polarity and charges up the capacitor C via the diode D. Energy is transferred to the output during the off-time of the transistor and this is known as the flyback time.

During the flyback time, current ramps down in the inductor, i.e.

$$V_0 = Ldi/dt$$

$$di = (V_0/L) \, dt$$

At the end of the flyback time,

$$I_1 - I_0 = (V_0/L)t_f \tag{7.5}$$

Equating (7.4) to (7.5):

$$V_1 t_n = V_0 t_f$$

$$V_0 = V_1 t_n/t_f = V_1 \, DT/(1-D) \, T$$

$$\boxed{V_0 = V_1 \, D/(1-D)} \tag{7.6}$$

Hence, with $D < 0.5$, the converter will step-down, and with $D > 0.5$ the converter will step-up.

The converter output voltage is a polarity reversal of the input voltage.

Example 7.2 _____

A buck–boost converter, as shown in Fig. 7.7, has an 18 V d.c. input, a switching frequency of 40 kHz, and an inductor of 50 μH. The capacitor value is 100 μF and the load resistor is 3 Ω.

Determine the output voltage, load current, and the average, maximum and minimum currents in the inductor for:

(a) on-time of 15 μs, and off-time of 10 μs;

(b) on-time of 10 μs, and off-time of 15 μs.

Solution

$$V_0 = V_1 \, D/(1-D)$$

$$T = 1/f = 1/40 \times 10^3 = 25 \, \mu s$$

$$D = t_n/T$$

(a) D $= 15/25 = 0.6$

V_0 $= V_1 D/(1-D)$

$= 18 \times 0.6/(1 - 0.6) = 27\,\text{V}$

Load current is

I_R $= V_0/R_L = 27/3 = 9\,\text{A}$

$I_1 - I_0 = (V_1/L)t_n$

$= (18/50) \times 15 = 5.4\,\text{A}$

Average inductor current is

I_L $= V_0 I_R/V_1$

$= 27 \times 9/18 = 13.5\,\text{A}$

$I_1 - I_0 = (V_1/L)t_n$

$= (18/50) \times 15 = 5.4\,\text{A}$

$I_1 = 13.5 + (5.4/2) = 16.2\,\text{A}$

$I_0 = 13.5 - (5.4/2) = 10.8\,\text{A}$

(b) D $= 10/25 = 0.4$

V_0 $= V_1 D/(1-\text{D})$

$= 18 \times 0.4/(1 - 0.4) = 12\,\text{V}$

Load current is

I_R $= V_0/R_L = 12/3 = 4\,\text{A}$

Average inductor current is

I_L $= V_0 I_R/V_1$

$= 12 \times 4/18 = 2.67\,\text{A}$

$I_1 - I_0 = (V_1/L)\,t_n$

$= (18/50) \times 10 = 3.6\,\text{A}$

$I_1 = 2.67 + (3.6/2) = 4.47\,\text{A}$

$I_0 = 2.67 - (3.6/2) = 0.87\,\text{A}$

7.4   ISOLATED BUCK–BOOST CONVERTER

To achieve isolation, the inductor in Fig. 7.7 is replaced by the magnetizing inductance of an isolating transformer. This can also be used to step up the output voltage, or as a means of producing multiple outputs.

One possible arrangement is shown in Fig. 7.8.

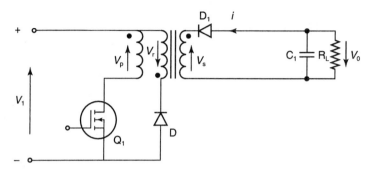

Figure 7.8

With Q_1 switched on, current flows in the primary, and energy is stored in the transformer inductance. D_1 is reverse-biased and no secondary current flows. Q_1 turns off, the primary inductive voltage changes polarity to keep the current flowing, secondary induced voltage changes polarity, and diode D_1 is forward-biased causing capacitor C_1 to charge up to the polarity shown. Meanwhile, the reset winding voltage V_r reverses and exceeds V_1, diode D_2 is forward-biased allowing the transformer core to reset via winding N_r as energy is returned to the d.c. source.

The output voltage is that of the buck–boost converter with the addition of the transformer turns ratio.

Output voltage is

$$V_0 = V_1 \, (N_s/N_p) \, D/(1-D)$$

(7.7)

7.5   PUSH–PULL CONVERTER

The circuit of a push–pull converter is shown in Fig. 7.9. Transistors Q_1 and Q_2 are switched on alternately. With Q_1 switched on, transformer action forward biases diode D_2 and energy is transferred to the output circuit as the inductor current ramps up. With Q_2 switched on, diode D_1 is forward-biased, the inductor current again ramps up and energy is transferred to the output circuit. The full transformer primary voltage appears across the switch that is off, i.e. $2V_1$. The output voltage is twice that of the forward converter, since there is no separate inductor charging period, and addition of the transformer turns ratio gives the output voltage as:

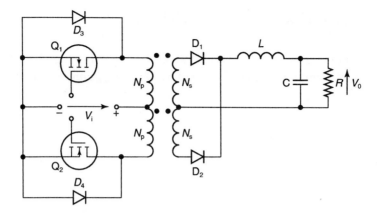

Figure 7.9

$$V_0 = 2D \, V_1 \, (N_s/N_p)$$

(7.8)

7.6 HALF-BRIDGE CONVERTER

The circuit diagram of the half-bridge converter is given in Fig. 7.10. Capacitors C_1 and C_2 each have half the d.c. input voltage across them, and this voltage, $V_1/2$, is switched across the primary winding by transistors Q_1 and Q_2 alternately. The switch that is off will have the full d.c. voltage across it. The output voltage expression is given by

$$V_0 = V_1 \, D \, (N_s/N_p)$$

(7.9)

The anti-parallel diodes shown across the transistors are those associated with all Mosfet switches, i.e. built-in parasitic diodes.

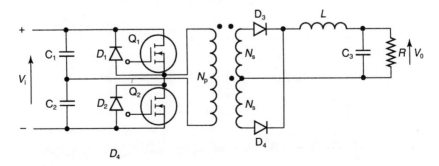

Figure 7.10

7.7 FULL-BRIDGE CONVERTER

The circuit diagram of the full-bridge converter is given in Fig. 7.11. Q_1 and Q_3 are switched on simultaneously, followed by Q_2 and Q_4 simultaneously. At least two of the bridge transistors require gate isolation circuits. The output voltage is twice that of the half-bridge:

$$V_0 = 2 V_1 D (N_s/N_p)$$ (7.10)

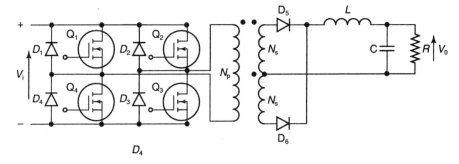

Figure 7.11

7.8 ANALYSIS OF THE PUSH–PULL CONVERTER

(a) In the circuit of the push–pull converter shown in Fig. 7.12, transistor Q_1 is switched on and transistor Q_2 is switched off; diode D_1 is reverse-biased and diode D_2 is forward-biased.

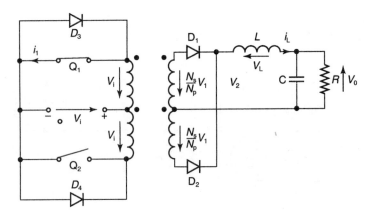

Figure 7.12

$$V_2 = (N_s/N_p) V_1$$

Inductor voltage is

$V_L = V_2 - V_0$

$\quad = (N_s/N_p) V_1 - V_0$

$\quad = L \, \Delta i_L/\Delta t$

Now turn-on time $\Delta t = DT$; hence change of inductor current is given by

$$\Delta i_L = (\Delta t/L) (V_2 - V_0) = (DT/L) ((V_1 N_s/N_p) - V_0)) \qquad (7.11)$$

(b) Q_1 is turned off, leaving both switches off. The circuit, showing the secondary side only, is given in Fig. 7.13.

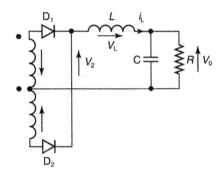

Figure 7.13

When Q_1 is switched off, the inductor voltage V_L changes polarity to keep the current flowing. Both diodes, D_1 and D_2, are then forward-biased and each carries half the inductor current. The inductor current ramps down. Overall secondary voltage is zero and V_2 is zero:

$V_L - V_0 = 0$

$L \, \Delta i_L/\Delta t = V_0$

$\Delta i_L = \Delta t \, V_0/L$

Off-time is

$\Delta t = (T/2) - DT = T (0.5 - D)$

$$\Delta i_L = T (0.5 - D) V_0/L \qquad (7.12)$$

Process (a) is repeated when Q_2 is switched on with Q_1 off, and process (b) when both switches are again off. For a symmetrical system, rise and fall currents are equal; equations (7.11) and (7.12) can be equated to give

$(DT/L)(V_1 N_s/N_p - (V_0)) = T (0.5 - D) V_0/L$

$DV_1 N_s/N_p - DV_0 = 0.5V_0 - DV_0$

Hence

$$V_0 = 2D \ V_1 \ N_s/N_p$$ (7.13)

Waveforms of transistor and inductor currents are shown in Fig. 7.14.

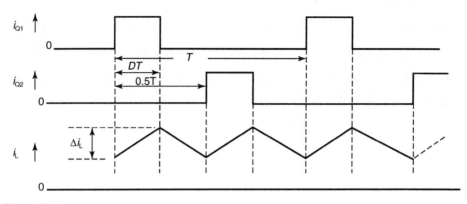

Figure 7.14

7.9   SELECTION OF INDUCTOR AND CAPACITOR VALUES

During the off-time of both switches, the change of inductor current is given by equation (7.12) as

$$\Delta i_L = T \ (0.5 - D) \ V_0/L$$

$$= (0.5 - D) \ V_0/fL$$

Half of this value must not exceed average inductor current, or load current, for continous current operation. Therefore, average load current is

$$I_L = V_0/R$$

$$= (0.5 - D) \ V_0/2 \ fL_{min}$$

$$L_{min} = R \ (0.5 - D)/2f$$ (7.14)

The inductor current ripple frequency is twice the converter frequency (see Fig. 7.15). This ripple current will flow almost entirely through the low impedance of the capacitor.

Increase of charge is equal to the area under the current/time curve, i.e.

$$\Delta Q = 0.25 \ \Delta i_L \ (T/4)$$

$$= 0.25 \ T \ (0.5 - D) \ V_0 \ T/4L$$

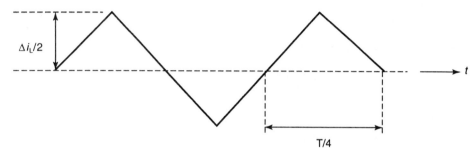

T/4

Figure 7.15

$$= (0.5 - D) \, V_0/16f^2 \, L$$

$$= (1 - 2D) \, V_0/32f^2 \, L$$

Now

$$\Delta Q = C \, \Delta V_0$$

where ΔV_0 is the capacitor ripple voltage. Hence

$$\boxed{C = \Delta Q/\Delta V_0 = (1 - 2D)/32 \, f^2 \, L(\Delta V_0/V_0)} \tag{7.15}$$

$(\Delta V_0/V_0) \times 100$ is the percentage ripple.

Example 7.3 _____

A push–pull, isolated SMPS is to deliver power into a 4Ω resistive load. The d.c. input voltage is 120 V, converter frequency is 20 kHz, transformer turns-ratio is unity and the duty cycle $D = 0.2$

(a) Determine suitable values for the converter inductor and capacitor, assuming continuous inductor current and 1% ripple of the output voltage.

(b) Using the values of L and C derived in (a), determine the maximum and minimum inductor currents.

(c) Repeat the calculations of (b) for a duty cycle of $D = 0.4$.

(d) Confirm that the minimum inductor value derived in (a) will result in a minimum inductor current of 0(A).

Solution

(a) V_0 $= 2DV_1 \, N_s/N_p = 2 \times 0.2 \times 120 \times 1 = 48 \, V$

   L_{min} $= R \, (0.5 - D)/2f$

   $= 4(0.5 - 0.2)/2 \times 20 \times 10^3 = 30 \, \mu H$

Choose $50 \mu H$.

$$C \quad = (1 - 2D)/32 \, f^2 \, L(\Delta V_0/V_0)$$
$$= (1 - 0.4)/32 \times 400 \times 10^6 \times 50 \times 10^{-6} \times 10^{-2} = 93.7 \mu F$$

Choose $100 \mu F$.

(b) Average inductor current is

$$I_L \quad = V_0/R = 48/4 = 12 \, A$$

$$\Delta i_L \quad = (0.5 - D) \, V_0/fL$$
$$= (0.5 - 0.2) \, 48/20 \times 10^3 \times 50 \times 10^{-6} = 14.4 \, A$$

$$i_{L(max)} = I_L + \Delta i_L/2 = 12 + 7.2 = 19.2 \, A$$

$$i_{L(min)} = I_L - \Delta i_L/2 = 12 - 7.2 = 4.8 \, A$$

(c) $V_0 \quad = 2DV_1 \, N_s/N_p$

$$= 2 \times 0.4 \times 120 \times 1 = 96 \, V$$

Average inductor current is

$$I_L \quad = V_0/R = 96/4 = 24 \, A$$

$$\Delta i_L \quad = (0.5 - D) \, V_0/fL = (0.5 - 0.4) \, 96/20 \times 10^3 \times 50 \times 10^{-6} = 6 \, A$$

$$i_{L(max)} = I_L + \Delta i_L/2 = 24 + 3 = 27 \, A$$

$$i_{L(min)} = I_L - \Delta i_L/2 = 24 - 3 = 21 \, A$$

(d) With $L_{min} = 30mH$

$$\Delta i_L \quad = (0.5 - D) \, V_0/fL$$
$$= (0.5 - 0.2) \, 48/20 \times 10^3 \times 30 \times 10^{-6} = 24 \, A$$

$$i_{L(min)} = I_L - \Delta i_L/2 = 12 - (24/2) = 0 \, A$$

Example 7.4 _____

A push–pull converter is to deliver 600W at 30V into a resistive load. The duty cycle is to be 0.4. Transformer turns ratio $N_s: N_p = 1: 4$. Determine suitable ratings for the Mosfet transistors.

Solution

$$I_L \quad = P/V_0 = 600/30 = 20 \, A$$

$$V_0 \quad = 2DV_1 \, N_s/N_p$$

$$V_1 = V_0 N_p/N_s\, 2D$$

$$= 30 \times 4/2 \times 0.4 = 150\,\text{V}$$

$$V_1 I_1 = V_0 I_0$$

$$I_1 = 30 \times 20/150 = 4\,\text{A}$$

Figure 7.16 shows the shape of the transistor current waveform.

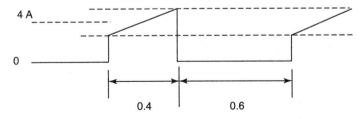

Figure 7.16

Transistor rms current is

$$I_{rms} = I_1 \sqrt{0.4} = 0.632 I_1 = 0.632 \times 4 = 2.53\,\text{A}$$

Off-state voltage $= 2\,V_1 = 2 \times 150 = 300\,\text{V}$

A 400 V, 5 A transistor with heat sinking would be suitable.

7.10 RESONANT INVERTER

In Section 4.8, a half-controlled bridge with a resonant load was analysed. The frequency was low, at 100 Hz, and the reactive components correspondingly large. The transistors switched at each half-cycle of the natural resonant frequency, and due to the damped nature of the circuit the switching occurs at significant initial conditions on the inductor and capacitor.

The higher the frequency of the inverter, the smaller the size of the inductive components. Unfortunately with the rectangular voltage waveforms, the higher the frequency, the higher the switching loss due to switch voltage and current being non-zero at the switching instant. Switching loss is directly proportional to frequency for these inverters because the switching loss occurs at each half-periodic time.

In resonant inverters, switching takes place at voltage or current zero, tending to eliminate the switching loss. For isolated inverters there is also the advantage that the transformer core is automatically reset.

7.11 SERIES LOADED RESONANT INVERTER

The circuit diagram of a series loaded inverter is shown in Fig. 7.17.

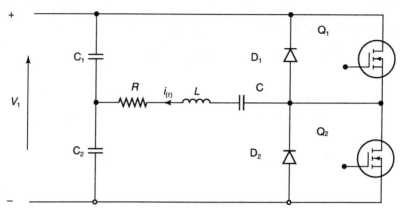

Figure 7.17

At each switch-on half the d.c. source voltage is applied to a series resonant circuit consisting of L, C and load resistance R.

Assumptions are that C_1 and C_2 are large enough not to affect the resonant frequency and that the switches are ideal.

Analysis of the inverter operation is done using the Laplace transformation and the circuit in the s domain. The circuit is underdamped, with $(R/2L)^2 < 1/LC$. Initial conditions at switching are zero.

The time-domain circuit at switching is shown in Fig. 7.18 (a), and the s domain circuit at switch closure is shown in Fig. 7.18 (b).

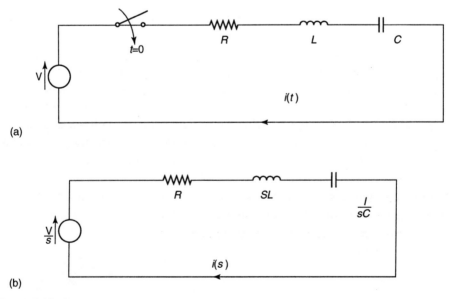

(a)

(b)

Figure 7.18

Now

$$i(s) = (V/s)/(R + sL + i/sC)$$

$$= V/(sR + s^2L + 1/C)$$

$$= (V/L)/(s^2 + s(R/L) + 1/LC)$$

$$= (V/L)/(s + R/2L)^2 + ((1/LC) - (R/2L)^2) \tag{7.16}$$

From a table of Laplace transform-pairs, the time domain solution is

$$\boxed{i(t) = (V/\omega L) \exp(-at) \sin \omega t} \tag{7.17}$$

where $a = R/2L$ and $\omega^2 = (1/LC) - (R/2L)^2$.

For the undamped circuit with $R = 0$, equation (7.16) becomes

$$i(s) = (V/L)/((s)^2 + (1/LC))$$

$$= (V/\omega L) \omega/((s)^2 + (\omega)^2)$$

$$i(t) = (V/\omega L) \sin \omega t$$

Inductor voltage is

$$v_L(s) = i(s) \times sL = V (s/(s^2 + \omega^2))$$

$$v_L = V \cos \omega t$$

Capacitor voltage is

$$v_C (s) = i(s) \times 1/sC = V(\omega/s(s^2 + \omega^2))$$

$$v_C(t) = V(1 - \cos\omega t)$$

At the end of one complete cycle, the capacitor voltage is zero in the undamped case.

At the end of one complete cycle, the capacitor voltage would be zero in the undamped case of a normal R–L–C resonant circuit, but with the diode in circuit, in the negative half-cycle current ceases when the diode is no longer forward-biased and this leaves an initial condition on the capacitor. This means that symmetrical current waveform takes a cycle or so to establish itself.

Let the natural frequency of the series resonant circuit be defined as $\omega_n = \sqrt{1/LC}$ for the undamped case, and $\omega_n = \sqrt{(1/LC - R^2/4L^2)}$ for the underdamped case. The on-time of each transistor is taken to be half the periodic time of this natural frequency.

The switching frequency, ω_s, of the transistors is chosen to be lower than ω_n, resulting in a longer periodic time for the switching frequency than that for the natural frequency, i.e.

$$\omega_s < \omega_n \quad \text{and} \quad T_s > T_n$$

The load current value can be altered by variation of the switching frequency.

Considering the circuit in Fig 7.17, with Q_1 switched on the positive half-cycle of current flows right to left through the load. The capacitor C charges to its maximum value by the end of the transistor on-time, approximately the d.c voltage V_1; at this time the current has fallen to zero. The negative half-cycle starts automatically, since as Q_1 turns off, D_1 is forward-biased and the capacitor C circulates current left to right through the load. At the end of the negative half-cycle, current ceases and diode D_1 turns off. An initial condition exists on the capacitor C when the diode turns off. After a small delay, Q_2 is turned on to start the load current cycle in reverse. The typical load current waveform is shown in Fig. 7.19.

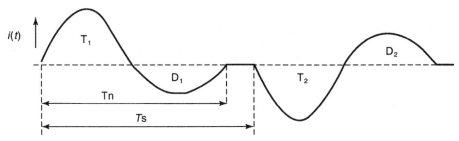

Figure 7.19

Example 7.5 _____

A half-bridge inverter has the circuit diagram shown in Fig. 7.17: $V_1 = 100\,\text{V}$, $L = 100\,\mu\text{H}$, $C = 0.1\,\mu\text{F}$ and $R = 10\,\Omega$.

(a) Determine the maximum current in the first positive half-cycle assuming the circuit is undamped.

(b) Is the circuit undamped? If not what kind of damping does it have? Repeat the calculation for (a) on the underdamped circuit.

Solution

(a) $i(t)$ $= (V/\omega_n L) \sin \omega t$

  ω_n^2 $= 1/LC = 1/10^{11}$

  $\therefore \omega_n$ $= 3.16 \times 10^5 \text{ rad/s}$

  $V/\omega_n L$ $= 50/3.16 \times 10^5 \times 10^{-5} = 1.58$

  $i(t)$ $= 1.58 \sin (3.16 \times 10^5) \text{ A}$

Maximum current is 1.58 A.

(b)   For the underdamped circuit $(R/2L)^2 < 1/ LC$. In this circuit

  $(R/2L)^2 = (10/2 \times 10^{-4})^2 = 25 \times 10^8$

  $1/LC$ $= 1/(10^{-4} \times 10^{-7}) = 10^{11}$

Hence the circuit is underdamped and $i(t) = (V/\omega L) \exp(-at) \sin \omega t$. Finding the maximum value of this is a little more difficult. The expression must be differentiated and then equated to zero to find the time at which maximum current occurs. This time is insered in the current expression to find the maximum current.

Differentiating the current expression and setting to zero gives

$\delta i/\delta t = V/\omega L \{\exp(-at) \omega \cos \omega t + \sin \omega t(-a \exp(-at)\} = 0$

$\therefore \exp(-at) \omega \cos \omega t = \sin \omega t (a \exp{-at})$

Hence

$\tan \omega t = \omega/a$

Now $a = R/2L = 5 \times 10^{-4}$ and $\omega = 3.122 \times 10^5$ rad/s. Thus

$\omega/a \quad = 6.244$

$\tan^{-1} (6.244) = 1.412$ rad $= \omega t$

$t = 1.412/3.122 \times 10^5 = 4.52 \mu s.$

This is the time for maximum current.

$at = 5 \times 10^4 \times 4.52 \times 10^{-6} = 0.266$

$\exp(-at) = 0.798$

$\sin \omega t = \sin (3.122 \times 10^5 \times 4.52 \times 10^{-6}) = \sin (1.41) = 0.987$

$i(t)_{max} = (V/\omega L) \exp(-at) \sin \omega t = 1.58 \times 0.798 \times 0.987 = 1.244 \, A$

7.12 PARALLEL LOADED RESONANT INVERTER

The circuit diagram of a parallel loaded resonant inverter is shown in Fig.7.20(a); the effective circuit at first switch-on is shown in the s domain form in Fig 7.20(b). This assumes ideal switches.

Norton's theorem is a convenient method of analysis for this type of circuit, and the Norton equivalent circuit is shown in Fig. 7.20(c). In Fig 7.20(b) the load short-circuit current is

$I_{sc} = V/s^2 L$

The internal impedance is

$Z_{int} = sL/sC (sL + 1/sC) = sL/(s^2 LC + 1)$

$i(s) = I_{sc} Z_{int}/(Z_{int} + R)$

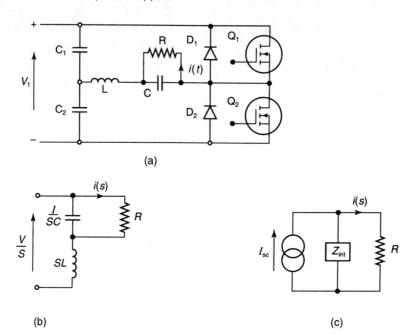

(a)

(b)   (c)

Figure 7.20

This simplifies to give

$$i(s) = V/LCRs \, (s^2 + s/CR + 1/LC) \tag{7.18}$$

A table of Laplace transform pairs brings this back to the time-domain as

$$i(t) = (V/R) \, \{1 - \exp(-\zeta\omega_n t)(\cos \omega t + (\zeta\omega_n/\omega) \sin \omega t))\} \tag{7.19}$$

where

$$\zeta \; = (1/2R) \, (\sqrt{L/C})$$
$$\omega_n = \sqrt{1/LC}$$
$$\omega \; = \omega_n \sqrt{1 - \zeta^2}$$

and

$$\zeta < 1$$

Example 7.6 ⎯⎯⎯⎯⎯⎯⎯⎯⎯⎯⎯⎯⎯⎯⎯⎯⎯⎯

The parallel loaded inverter in Fig. 7.20(a) has a d.c. input voltage of 60 V, and resonant circuit components of $L = 100 \; \mu H$ and $C = 1 \, \mu F$. The load resistance R = 20Ω.

Determine an expression for the load current at switching, assuming zero initial conditions on the inductor and capacitor, and hence find the load current value 15 μs after switching.

Solution

$$\zeta \quad = (1/2R) (\sqrt{L/C})$$

$$= (1/40) (\sqrt{10^{-4}/10^{-6}} = 0.25$$

$$\omega_n \quad = \sqrt{1/LC} = \sqrt{1/10^{-4} \times 10^{-6}} = 10^5 \text{ rad/s}$$

$$f_n \quad = (1/2\pi) \, 10^5 = 15.9 \, \text{kHz}$$

$$\omega \quad = \omega_n \sqrt{1 - 0.25^2} = 0.97 \times 10^5 \text{ rad/s}$$

$$\zeta\omega_n \quad = 0.25 \times 10^5$$

At $t = 15 \, \mu s$

$$\zeta\omega_n t \quad = 0.25 \times 10^5 \times 15 \times 10^{-6} = 0.375$$

$$\cos\omega t \quad = \cos (0.97 \times 10^5 \times 15 \times 10^{-6})$$

$$= \cos (1.455) = 0.116$$

$$\sin \omega t = \sin (1.455) = 0.993$$

$$\zeta\omega_n/\omega = 0.25 \times 10^5/0.97 \times 10^5 = 0.257$$

$$V \quad = V_{dc}/2 = 30$$

$$V/R \quad = 30/20 = 1.5 \, \text{A}$$

Substituting values in equation (7.19):

$$i(t) \quad = 1.5\{1 - 0.687(0.116 + (0.257 \times 0.993))\} = 1.12 \, \text{A}$$

The resonant converters covered so far can be readily analysed using simulation methods, such as PSPICE and MICROCAP. The waveforms, taking initial conditions into account, are easily plotted, as are the rms values of the voltages and currents. The simulation methods give rapid solutions to these complex circuits. However, they do not replace the need to use analytical methods to verify the simulation results.

The resonant inverters discussed in Sections 7.11 and 7.12 can be used as part of an isolated switched mode power supply. The addition of a transformer, a rectifier and a smoothing capacitor as shown in Fig. 7.21 will produce either a fixed d.c. output voltage with fixed resonant and switching frequencies, or variable output voltage controlled by switching frequency and transformer turns ratio. The bridge can be half or full, with the full-bridge giving twice the output voltage of the half-bridge, but with the extra complexity of the base drive circuit.

Consider the circuit in Fig. 7.21. The referred load resistance is

$$R' = R \times (N_p/N_s)^2$$

$$i(t) = (V_1/2R') \{1 - \exp(-\zeta\omega_n t)(\cos \omega t + (\zeta\omega_n/\omega) \sin \omega t)\}$$

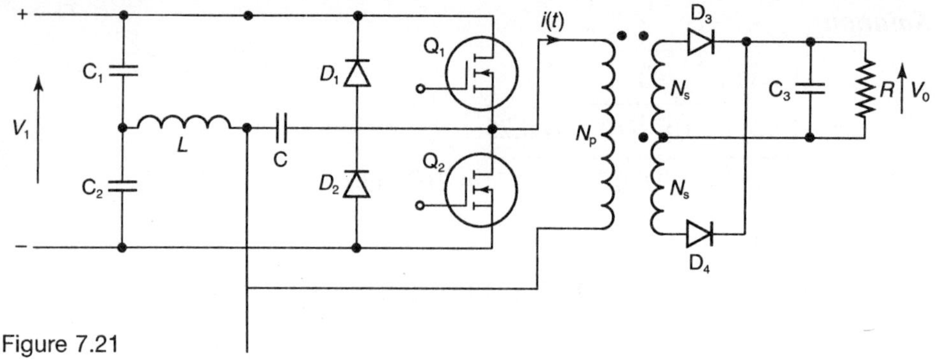

Figure 7.21

Depending on the size of the filter capacitor, C_3, and neglecting device voltage drops,

$$V_0 \cong i(t)_{\text{max}} \, R'(N_s/N_p)$$

7.13 UNINTERRUPTIBLE POWER SUPPLIES

Uninterruptible power supplies (UPSs) and standby power supply systems are used in applications where loss of the mains supply could be disastrous, as in the case of hospital operating theatres or intensive care units, computer installations, production systems, alarms and signalling equipment.

The UPS can be on-line or off-line. Both systems use a d.c. link inverter with a battery bank and trickle-charger. In the case of the off-line system, in normal operation power is supplied directly from a.c. mains. In the event of mains failure, a transfer switch disconnects the power line and connects the inverter to the load. When mains power is restored, the load is reconnected to the power line. A block schematic diagram of the off-line system is given in Fig. 7.22. The switching process can take several milliseconds if the switch is solid state, and tens of milliseconds if the switch is electromechanical.

With on-line systems, the rectifier–inverter combination supplies the load power from the a.c. mains during normal operation. Should the mains fail, the battery atuomatically supplies the d.c. link to the inverter and there is no time delay involved. Should the rectifier–inverter system fail, the load could be transferred to a.c. mains using a transfer switch.

7.14 SELF-ASSESSMENT TEST

1 What are the advantages of using SMPSs over linear regulators?

2 (a) Why are transformers used in SMPSs?

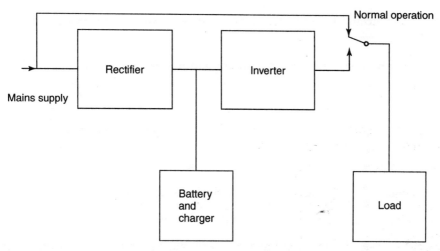

Figure 7.22

(b) Why is it necessary to reset the transformer core?

(c) Describe two ways of resetting the core in isolated converters.

3 (a) Sketch the circuit diagram of a buck–boost converter and state the output voltage equation.

(b) What is meant by the term buck–boost?

4 State the output voltage equations for isolated converters that are (a) push–pull, (b) half-bridge, (c) full-bridge.

5 Sketch waveforms of transistor and inductor currents against time for a push–pull converter operating in the continuous current mode.

6 Describe the essential difference between forward and flyback converters.

7.15 PROBLEMS

1 An isolated buck converter has a turns ratio of $N_s{:}N_p = 1{:}8$, an output voltage of 5 V, and a load resistance of $0.2\,\Omega$.

(a) Determine the output power, load current and duty cycle for an input voltage of $100\,V$ d.c.

(b) The converter has been designed to use the following values: $f = 40\,kHz$, $L = 30\,\mu H$, $C = 50\,\mu F$. Calculate the turn-on and turn-off times of the transistor and the variation of inductor current.

2 An isolated buck-boost converter is to supply $500\,W$ at $48\,V$ into a resistive load. The d.c. input voltage is $24\,V$, and the transformer turns–ratio is unity. The inductor value is $100\,\mu H$ and the converter frequency is $20\,kHz$.

(a) Calculate the value of the duty-cycle required and the resistance of the load.

(b) Determine the average, maximum and minimum values of the inductor current.

3 A push–pull isolated SMPS provides d.c. power to a resistive load at 30 V, 20 A. The frequency of the converter is set to 10 kHz and the duty-cycle to $D = 0.2$. The transformer turns-ratio is $N_s:N_p = 1:2$. Output voltage ripple is not to exceed 2%

(a) Suggest suitable values for the inductor and capacitor.

(b) Calculate average, maximum and minimum values of inductor currents.

(c) Determine the input voltage required and the magnitude of the output voltage ripple.

8 Power electronic switches

8.1 INTRODUCTION

The power diode is used in a.c.–d.c. rectification circuits where fixed voltage d.c. is required. The diode conducts, or switches on, once the anode polarity is positive with respect to the cathode. When the anode polarity is negative with respect to the cathode, the diode ceases to conduct, or switches off. The diode has the highest rating of all the semiconductor switches and is the cheapest, but it cannot regulate the magnitude of the rectified voltage.

Thyristors require anode polarity positive with respect to the cathode to switch-on, but will not conduct until a low voltage, short-duration pulse is applied to the gate with gate polarity positive with respect to the cathode. For a.c. supply applications, this method of voltage magnitude control blanks out part of one half-cycle of the mains voltage; the switch turns off at mains voltage zero and will not conduct in the other half-cycle as reverse polarity appears across it. The circuit is a half-wave controlled rectifier, and the method is known as phase-control. For full-wave controlled rectification, two thyristors are connected in inverse parallel, or a Triac can be used if the required rating is available. Thyristors and Triacs are ideal for a.c. applications since turn off occurs naturally at mains voltage zero for resistive loads, and current zero for inductive loads. For d.c. applications a forced-commutation, or turn-off, circuit is required to force the current to zero at switch-off.

Power bipolar junction transistors (BJTs) are used for power control where there is a d.c. supply available, as in the case of choppers and inverters. In the case of an n–p–n transistor, with the collector polarity positive with respect to the emitter, the transistor is turned on with current pulse on the base, making the base positive with respect to the emitter. This base pulse must remain on for the duration of the on-time of the transistor. For large power transistors, the base current could be 10 A or so. If a Darlington transistor is available with the required power rating, the base current would be reduced to about the 0.5 A level, at the cost of increased on-state voltage drop. The BJT cannot withstand a reversal of voltage across the collector–emitter terminals, and unless precautions were taken this could destroy the switch. Hence the BJT is not used with a.c. voltage across it. If the application requires the switch voltage to reverse, as in the case of inductive loads, an inverse-parallel diode (the feedback diode) must be connected across the transistor.

Power Mosfets, like BJTs, have no reverse voltage withstand capability, and are therefore suited to switching d.c. power sources but not a.c. sources. The Mosfet has a built-in parasitic or body diode. This will conduct if the voltage across the transistor reverses. For normal use, the drain terminal is of positive polarity with respect to the source terminal. A voltage pulse on the gate terminal will turn the transistor on, a charge will flow into the gate, and once this is finished the gate current is negligible (of the order of nA). The voltage pulse must remain high for the duration of the transistor on-period. Speed of switch-on for the Mosfet is about 10 times faster than that for the BJT, and the low base current requirement makes it the obvious choice if the rating is available.

The insulated gate bipolar transistor (IGBT) combines the power handling capability of the BJT with the fast switching of the Mosfet. It requires a higher gate voltage drive and has a higher on-state voltage drop than the Mosfet, tending to rule it out of contention for low voltage applications.

Table 1.1 (p. 3) gives details of the comparative performance of power electronic switches. All the switches are layers, or sandwiches, of semiconductor p–n junctions. In all cases when the switch is off, only leakage current flows. When the switch is on, the current is limited by load.

Manufacturers of power electronic switches also produce integrated circuit modules to control the turn-on and turn-off of the power devices. These are known as gate and base drivers. There is also the continuing development of 'smart power devices' or 'intelligent power modules', in which the switch, the drive and the protection and control circuits are all contained in one module. Some of these manufacturers are listed below:

- Harris Semiconductors
- Hitachi
- International Rectifier
- Motorola Semiconductors
- Plessey
- Semicron
- Siliconex Incorporated.

These companies will readily provide data sheets and application notes on their products. Another source of information is the component distributor, such as Radio Spares and Farnell.

8.2 THYRISTORS

The circuit symbol, semiconductor arrangement, two-transistor analogy and two transistor symbol are shown in Fig. 8.1 (a)–(d).

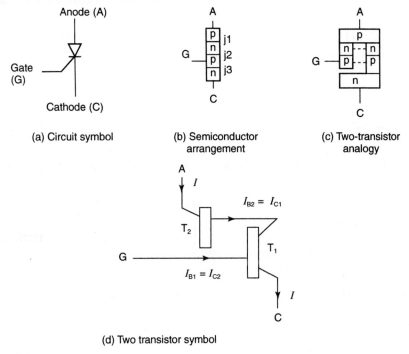

(a) Circuit symbol (b) Semiconductor arrangement (c) Two-transistor analogy

(d) Two transistor symbol

Figure 8.1

With normal forward bias, i.e. anode positive with respect to cathode, and no gate signal, junctions j1 and j3 are forward-biased and j2 is reverse-biased. Only leakage current flows through the thyristor. With reverse bias, i.e. cathode positive with respect to the anode, j2 is forward-biased and j1 and j3 are reverse-biased. Again only leakage current flows.

I_{co} = leakage current

$I_{c1} = a_1 I$

$I_{c2} = a_2 I$

$I = I_{co} + I_{c1} + I_{c2}$

$I = I_{co} + a_1 I + a_2 I$

$I(1 - a_1 - a_2) = I_{co}$

$$\boxed{I = I_{co}/1 - (a_1 + a_2)}$$ (8.1)

The value of a depends on the emitter current, it is very low when only leakage current flows. For $a_1 + a_2 \ll 1.0$, I, $\cong I_\infty$ and the thyristor is switched off.

Now let a short-duration gate pulse be applied; I_G will cause I_{B1} to increase. This will increase I_{C1} and put T_1 into the on-state. Increase of I_{C1} also means increase of I_{B2} and I_{C2} which turns on T_2.

The current through the thyristor is now only limited by the load, and the device is hard on. The loop between the two transistors is regenerative and is self-sustaining when I_G is removed.

As the emitter current increases, the value of $(\alpha_1 + \alpha_2)$ approaches unity. From equation (8.1), $I \to \infty$, but of course this is limited by the value of the external load.

8.3   THYRISTOR CHARACTERISTICS

The typical load current against thyristor voltage drop characteristic is shown in Fig. 8.2.

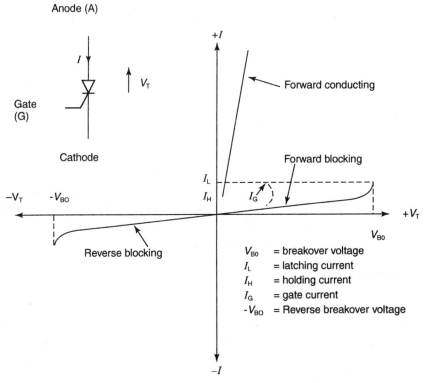

Figure 8.2

Transition from forward blocking to forward conducting occurs when a short-duration low voltage pulse producing the gating current I_G is applied to the gate. V_{BO} can also cause this transition, but it can also damage the device and is not used as a method of turn-on. I_L, the latching current through the thyristor, must be reached if the switch is to remain on. I_H is the holding current below which the current must fall if the thyristor is to turn-off.

$-V_{BO}$, the reverse breakover voltage, must be avoided.

8.4 THYRISTOR TURN-ON

The gate circuit

The larger the gate current and voltage, the faster the turn-on time of the thyristor. However the manufacturer's maximum gate ratings must not be exceeded.

A thyristor gate circuit is shown in Fig. 8.3 and a typical gate characteristic in Fig. 8.4.

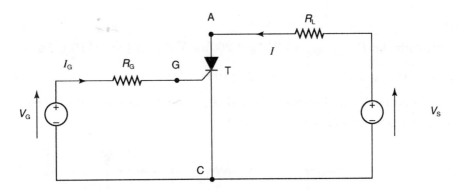

Figure 8.3

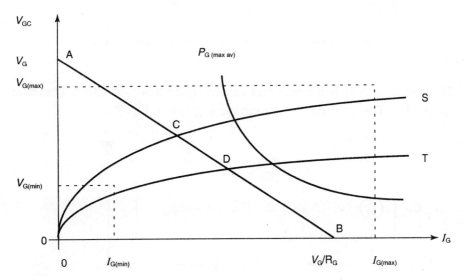

Figure 8.4

0S and 0T represent typical operating spread for the same type of thyristor.

The operating point on the gate characteristic will depend on the gate voltage and the gate resistance R_G (this is the sum of gate source and current limiting resistances). A load line, AB, is superimposed on the gate characteristic with a slope $1/R_G$, and this indicates the range of operating point lies between C and D.

The maximum gate power can be extended by pulse firing (as shown in Fig. 8.5). The time averaged maximum power is as follows:

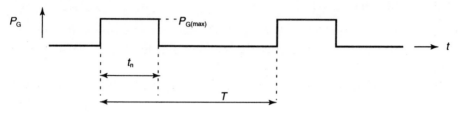

Figure 8.5

$$P_{G(\text{max av})} = P_{G(\text{max})} \, t_n/T = P_{G(\text{max})} \, t_n f$$

For example, let $P_{G(\text{max av})} = 2\,\text{W}$, $t_n = 200\,\mu\text{s}$, $T = 1\,\text{ms}$ ($f = 1\,\text{kHz}$). Then

$$P_{G(\text{max})} = P_{G(\text{max av})} \, T/t_n = 2 \times 1/0.2 = 10\,\text{W}$$

A gate firing circuit is shown in Fig. 8.6: R_1 is a gate current limiting resistor, and R_2 is a gate voltage limiting resistor.

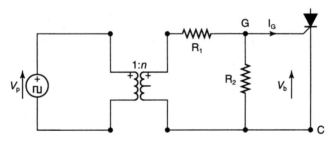

Figure 8.6

The gate circuit is non-linear and can easily be analysed using Thevenin's theorem. The Thevenin equivalent circuit is given in Fig. 8.7.

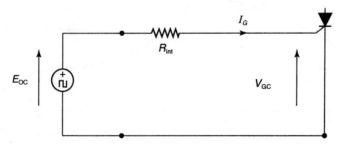

Figure 8.7

$E_{oc} = V_p \times n \times R_2/(R_2+R_1)$

$R_{int} = R_2 \times R_1 /(R_2+R_1)$

The graphical solution is shown in Fig. 8.8.

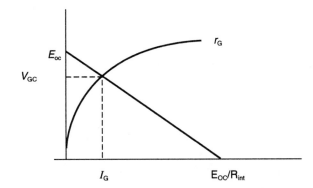

Figure 8.8

Example 8.1

A thyristor has a linearized gate-cathode characteristic of slope 25 V/A. A gate current of 200 mA turns the thyristor on in 16 μs. The gate source voltage is 10 V. The manufacturer's average maximum power for the gate is 400 mW. Pulse firing is used. Calculate:

(a) the value of the gate series resistance;

(b) the gate power dissipation during turn-on;

(c) the frequency of the gate pulses.

Solution

The gate resistance $r_g = V_{GC}/I_G = 25\,\Omega$.

(a) R_G (total) $= V_G/I_G = 10/0.2 = 50\,\Omega$

    Gate series resistance $= 50 - 25 = 25\,\Omega$

(b) Gate power dissipation $= I_G^2\, r_G = (0.2)^2 \times 25 = 1\,W$

(c) $f = P_{G(max\ av)}/P_G\, t_n = 0.4/1 \times 16 \times 10^{-6} = 25\,kHz$

Example 8.2

The range of spread of gate-cathode characteristics for a certain thyristor can be linearized to between 15 V/A and 10 V/A. The manufacturer's data gives the maximum gate power dissipation as 5 W.

Sketch the characteristic up to $V_{GC} = 15\,\text{V}$ and $I_G = 1.5\,\text{A}$, and insert the $P_{G(\text{max av})}$ line.

With the gate firing circuit as shown in Fig. 8.6, a 1:1 isolating transformer, V_p amplitude of 20 V, and $R_1 = R_2 = 20\,\Omega$, determine the possible range of V_{GC} and I_G.

Solution

The characteristic is sketched in Fig. 8.9.

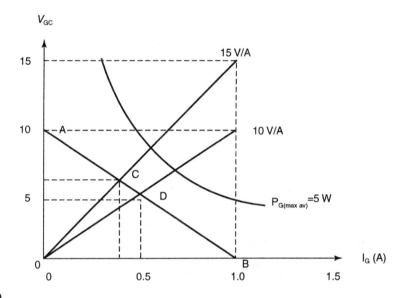

Figure 8.9

From Thevenin's theorem

$$V_{OC} = V_p \times R_1/(R_1 + R_2)$$

$$= 20 \times 20/(20 + 20) = 10\,\text{V}$$

$$R_{int} = R_1 R_2/(R_1 + R_2)$$

$$= 20 \times 20/40 = 10\,\Omega$$

Load line AB can be inserted.

This gives an operating region between C and D, i.e. about 5–7 V for V_{GC} and 0.4–0.5 A for I_G

Latching current

The turn-on time of a thyristor is defined as the time from the position where 10% of the gate current is established to the position where 90% of the anode current is established. In practice, the gate pulse must be on long enough for the anode latching current to be reached.

Gate power is wasted if the gate pulse is on longer than necessary. The manufacturer's data needs to be consulted to obtain successful firing.

If the load is inductive, the growth of anode, or forward, current will be delayed. This will mean either longer gate pulse duration or a train of short-duration pulses to ensure the latching current is obtained.

If the applied voltage is d.c., the growth of anode current is exponential given by

$$i(t) = (V/R) (1 - \exp(-Rt/L)) \tag{8.2}$$

At the instant of switching, the rate of current growth is given by differentiating equation (8.2), and setting $t = 0$, i.e.

$$di/dt = (V/L) \exp(-Rt/L)$$

$$(di/dt)_{t=0} = V/L \ (A/s) \tag{8.3}$$

Assume that the rate of growth of current is constant over the short duration of the gate pulse; the position is shown in Fig. 8.10.

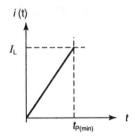

Figure 8.10

If $t_{P(min)}$ is the minimum duration of the gate pulse necessary to obtain the latching current, then comparing equation (8.3) with Fig. 8.10

$$(di/dt)_{t=0} = V/L = I_L/t_{P(min)}$$

$$t_{P(min)} = L \ I_L/V \tag{8.4}$$

For the case of a d.c. supply of $V = 25\,V$, a circuit inductance of 50 mH and a thyristor latching current of 50 mA, the minimum pulse length would be

$$t_{P(min)} = L \ I_L/V = 50 \times 10^{-3} \times 50 \times 10^{-3}/25 = 100\,\mu s$$

If the supply is sinusoidal a.c. with an R–L load, the time-varying current at switch-on is given by

$$i(t) \quad = (E_m/|Z|) \ \{\sin(\omega t + a - \phi) + \sin(\phi - a)\exp(-Rt/L)\}$$

$$di/dt \quad = (E_m/|Z|)\{\omega\cos(\omega t + a - \phi) - (R/L) \sin(\phi - a)\exp(-Rt/L)\}$$

At $t = 0$

$$(di/dt)_{t=0} = \omega(E_m/|Z|)\{\cos(a - \phi) - (R/\omega L) \sin(\phi - a)\} \tag{8.5}$$

Equation (8.5) can be used in the same way as equation (8.4) to find the minimum gate pulse width (see Example 8.3).

Example 8.3 _____

A thyristor has a latching current of 40mA. It is connected in a half-wave controlled rectifier circuit, between a 120V, 50Hz a.c. supply and an inductive load of 20 Ω resistance and 30Ω inductive reactance. Determine the minimum pulse width required to obtain latching current when the firing angle delay a is 30°.

Solution

From equation (8.5)

$$(di/dt)_{t=0} = \omega(E_m/|Z|)\{\cos(a - \phi) - (R/\omega L) \sin(\phi - a)\}$$

where

ω $\quad = 2\pi \times 50 = 314.2$ rad/s

E_m $\quad = \sqrt{2} \times 120 = 170$V

$|Z|$ $\quad = \sqrt{20^2 + 30^2} = 36\Omega$

ϕ $\quad = \arctan(30/20) = 56.3°$

$(di/dt)_{t=0} = 314.2(170/36)\{\cos(30° - 56.3°) - (20/30) \sin(56.3° - 30°)\}$

$\qquad\qquad = 1484(0.896 - 0.296) = 890$ A/s

Assuming di/dt is constant

$$(di/dt)_{t=0} = I_L /t_{P(min)} = 890$$

Therefore

$t_{P(min)}$ $\quad = I_L/890 = 40 \times 10^{-3}/890 = 45\mu s$

8.5 POWER MOSFETS

The semiconductor arrangement of an n-channel enhancement Mosfet is shown in Fig. 8.11 and the circuit diagram schematic in Fig. 8.12.

With the normal forward polarity for V_{DD} on the Mosfet, as shown in Figs 8.11 and 8.12, but with $V_{GS} = 0$, the device is like an npn transistor with the drain to gate junction reverse-biased, and therefore no drain current flow.

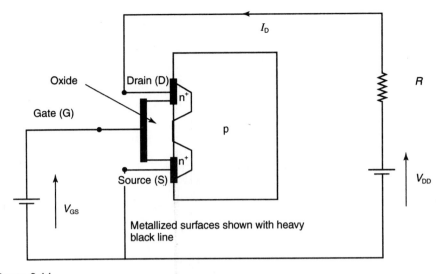

Figure 8.11

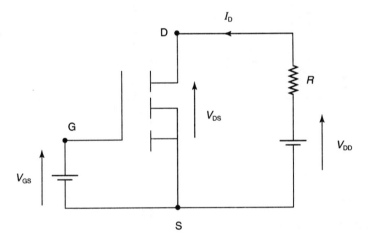

Figure 8.12

With V_{GS} applied, making the gate positive with respect to the source, positive charge accumulates at the gate metallic surface, an electric field is created in the oxide layer, and negative charge accumulates at the p-structure surface in contact with the oxide layer. This negative charge repels holes in the p-structure and leaves a virtual n-type channel through which electrons can flow from source to drain, i.e. conventional current flow from drain to source. For the Mosfet to turn on, V_{GS} must exceed the threshold voltage V_T.

The linearized transfer characteristic of the Mosfet is shown in Fig. 8.13, and the output, or drain-source, characteristic is shown in Fig. 8.14.

As with other power switches, the load line can be superimposed on the output characteristic to give the operating point, as shown in Fig. 8.14.

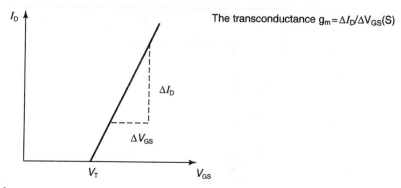

Figure 8.13

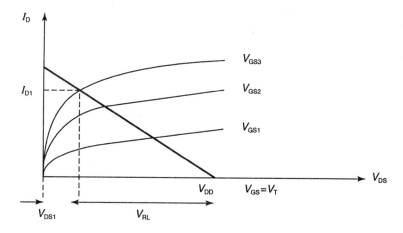

Figure 8.14

$$V_{DD} = I_D R + V_{DS}$$

$$I_D = (V_{DD}/R) - (V_{DS}/R)$$

At $I_D = 0$, $V_{DD} = V_{DS}$; at $V_{DS} = 0$, $I_D = V_{DD}/R$.

I_{D1} is the actual drain current, V_{DS1} is the Mosfet voltage drop and V_{RL} is the load voltage for gate source voltage V_{GS3}.

If the slope of the characteristic to the left of the intersection of the V_{GS} (working) curve with the load line, the so-called 'ohmic region', is linearized then a much simpler solution is obtained. The slope is referred to as the on-state resistance R_{DS}. The simple series equivalent circuit is shown in Fig 8.15.

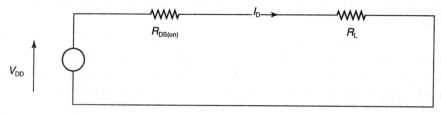

Figure 8.15

The switch manufacturers give typical values of $R_{DS(on)}$ for their products, e.g. a 400V, 5A device might have a $R_{DS(on)}$ of 1.0Ω, and a 200V, 25A device might have an $R_{DS(on)}$ value of 0.1Ω.

Example 8.4

An IRF150 power Mosfet has $V_{DD} = 20V$, $R_1 = 0.5\ \Omega$; at $V_{GS} = 8V$, the on-state resistance is 0.1Ω. Determine the values of load current, device voltage drop, load power and circuit efficiency.

Solution

From the circuit in Fig. 8.15

$$I_D \quad = \quad V_{DD}/(R_{DS(on)} + R_L)$$

$$= 20/(0.1 + 0.5) = 33.3A$$

$$V_{GS} = I_D \times R_{DS(on)} = 33.3 \times 0.1 = 3.33V$$

$$P_L \quad = (33.3)^2 \times 0.5 = 554W$$

$$P_{in} \quad = I_D^2 (R_{DS(on)} + R_L)$$

$$= (33.3)^2 \times (0.1 + 0.5) = 665W$$

Efficiency $= P_L/P_{in} = 554/665 = 0.833$ or 83.3%.

The graphical solution to Example 8.4 is shown in Fig. 8.16. With $V_{DS} = 0$

$$I_D = V_{DD}/R_L = 20 / 0.5 = 40A$$

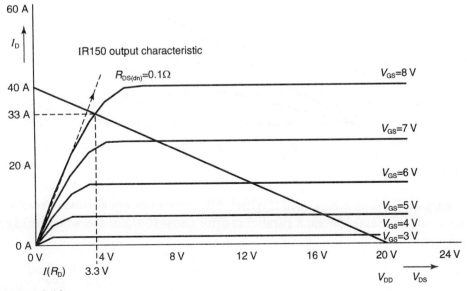

Figure 8.16

Superimposing the load line gives $I_D = 33\,A$ and $V_{DS} = 3.3\,V$, the same values as the equivalent circuit.

8.6   EXTENSION OF POWER SWITCH RATINGS

The ratings of power electronic switches can be extended by series connection for high voltage, and parallel connection for high current. Matching of switches as closely as possible will assist in equal sharing of voltage or current, but the normal range of production spread can be forced to share more equally using additional circuitry.

Series connection

Figure 8.17 shows series connected thyristors, but they could also be power transistors. Figure 8.18 shows the unequal sharing of blocking voltage in the off-state due to production spread.

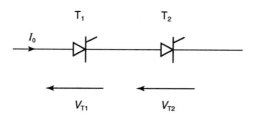

Figure 8.17

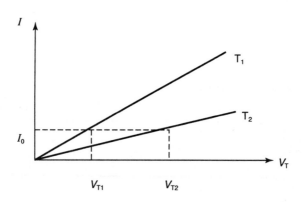

Figure 8.18

Parallel resistors across each thyristor will cause more equal voltage sharing. As a rule of thumb, let each parallel resistor carry 10 times the worst leakage current.

Example 8.5

Two 1200 V thyristors are series connected and 2 kV is connected across them. The forward blocking characteristic of the first thyristor has a slope of 16.7 μA/V, and the second thyristor a slope of 12.5 μA/V (see Fig. 8.19).

(a) How will they share the voltage?

(b) Voltage equalizing resistors of 6.8 kΩ are now connected across each thyristor. Calculate the voltage across each thyristor and the power dissipation in the equalizing resistors.

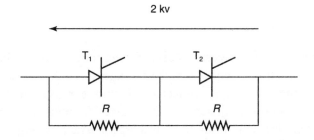

2 kv

Figure 8.19

Solution

(a) The inverse of the slope of the blocking characteristic is thyristor off-state resistance, R_T:

$$R_{T1} = 1/16.7 \times 10^{-6} = 60\,k\Omega$$

$$R_{T2} = 1/12.5 \times 10^{-6} = 80\,k\Omega$$

By voltage divider:

$$V_{T1} = V\,R_{T1}/R_{T1} + R_{T2}$$

$$= 2 \times 10^3 \times 60/140 = 857\,V$$

$$V_{T2} = V\,R_{T2}/R_{T1} + R_{T2}$$

$$= 2 \times 10^3 \times 80/140 = 1143\,V$$

(b) With the equalizing resistors connected, the effective thyristor resistances are now given by

$$R_{T1}' = 60 \times 6.8/(60 + 6.8) = 6.11\,k\Omega$$

$$R_{T2}' = 80 \times 6.8/(80 + 6.8) = 6.27\,k\Omega$$

$$V_{T1}' = V\,R_{T1}'/R_{T1}' + R_{T2}'$$

$$= 2 \times 10^3 \times 6.11/12.38 = 987\,V$$

$$V_{T2} = V R_{T2}'/R_{T1}' + R_{T2}'$$

$$= 2 \times 10^3 \times 6.27/12.38 = 1013\,\text{V}$$

Power dissipation in the first equalizing resistor is

$$(987)^2/6.8 \times 10^3 = 143\,\text{W}$$

and in the second resistor is

$$(1013)^2/6.8 \times 10^3 = 151\,\text{W}$$

Blocking state current is $2/(6.11 + 6.27) = 162\,\text{mA}$.

Dynamic equalization

During turn-off, the differences in stored charge, due to the different junction capacitances, will cause unequal reverse voltage sharing. The solution to this problem is to connect a parallel connected capacitor across each thyristor that is large enough to swamp the junction capacitance. A small resistance in series with this capacitance will limit the discharge current through the thyristor during switch-on. The R_2–C network will also act as a snubber network to limit the rate of rise of voltage across the thyristor at switch-on. The circuit arrangement for static and dynamic voltage equalization is shown in Fig. 8.20.

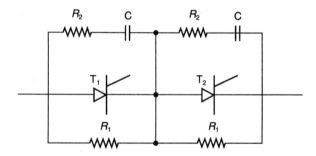

Figure 8.20

Reference books such as *Power Electronics* (Williams, 1992) have formalized the design of equalization networks, and give the values of R_1 and C as follows:

$$R_1 = (nV_D - V_S)/(n - 1)I_0 \quad \text{(ohm)} \tag{8.6}$$

$$C = (n-1)\,\Delta Q/(nV_D - V_S) \quad \text{(farad)} \tag{8.7}$$

where

n = the number of series connected thyristors

V_D = the voltage rating of each thyristor

V_S = the supply voltage

I_0 = the maximum leakage current

ΔQ = the maximum difference in recovery charge between thyristors of a particular type.

Example 8.6

Four 800 V thyristors are connected in series to share a 3 kV supply voltage. The maximum forward leakage current is 4 mA, and the maximum difference in recovery charge is 20 μC. The circuit arrangement for static and dynamic voltage sharing is that shown in Fig. 8.20 with $R_2 = 20\,\Omega$.

(a) Determine the values of R_1 and C, and the maximum discharge current, assuming equal voltage sharing.

(b) Repeat the calculations in (a) for the case where an extra series connected thyristor is used as a safety, or derating, factor.

Solution

(a) n = 4:

$$R_1 = (nV_D - V_S)/(n - 1)I_0$$

$$= (4 \times 800 - 3000)/(4 - 1) \times 4 \times 10^{-3}$$

$$= 16.7\,k\Omega$$

$$C = (n-1)\,\Delta Q/(nV_D - V_S)$$

$$= (4-1) \times 20 \times 10^{-6}/(3200 - 3000)$$

$$= 0.3\,\mu F$$

At switch-on

discharge current = thyristor voltage/R_2

$$= (3000/4 \times 20) = 37.5\,A$$

(b) n = 5:

$$R_1 = (nV_D - V_S)/(n - 1)I_0$$

$$= (5 \times 800 - 3000)/(5 - 1) \times 4 \times 10^{-3}$$

$$= 62.5\,k\Omega$$

$$C = (n-1)\,\Delta Q/(nV_D - V_S)$$

$$= (5-1) \times 20 \times 10^{-6}/(4000 - 3000\,)$$

$$= 0.08\,\mu F$$

At switch-on

discharge current = thyristor voltage/R_2

$$= (3000/5 \times 20) = 30\,\text{A}$$

Example 8.7

The thyristors used in Example 8.6 have maximum permitted rate of voltage and current change as shown below:

$(\delta v/\delta t)_{max} = 200\,\text{V}/\mu\text{s}$ and $(\delta i/\delta t)_{max} = 100\,\text{A}/\mu\text{s}$

Are the thyristors adequately protected at switch-on?

Solution

Capacitor discharge voltage at switch-on is given by

$v_C \qquad = V \exp(-t/CR_2)$

$(\delta v/\delta t) = (-V/CR_2) \exp(-t/CR_2)$

This is maximum at $t = 0$, i.e.

$(\delta v/\delta t)_{max} = -V/CR_2$

$$= -750/0.3 \times 10^{-6} \times 20 = -125\ \text{V}/\mu\text{s}$$

The above assumes equal voltage sharing.

Capacitor discharge current at switch-on is given by

$i \qquad = -(V/R_2) \exp(-t/CR_2)$

$(\delta i/\delta t) = (V/CR_2^2) \exp(-t/CR_2)$

This is maximum at $t = 0$, i.e.

$(\delta i/\delta t)_{max} = V/CR_2^2$

$$= 750/0.3 \times 10^{-6} \times 20^2 = 6.25\ \text{A}/\mu\text{s}$$

The thyristors are adequately protected.

Parallel connection

When thyristors are connected in parallel, they do not carry equal currents due to differences in their on-state characteristics. The device carrying the highest current will dissipate more power, causing heat increase, which in turn will reduce the on-state resistance, causing more current to flow; the imbalance is increased and could result in switch failure. Common heat-sinking of paralleled devices will help to equalize temperatures.

Static equalization of currents is improved by connecting low value series resistors. Dynamic equalization can be achieved by the use of magnetically coupled coils.

Figure 8.21 shows the way in which unequal currents flow before static equalization and Fig. 8.22 the circuit arrangement to implement static equalization.

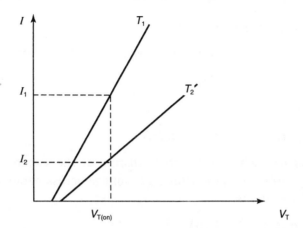

Figure 8.21

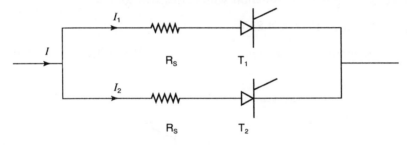

Figure 8.22

Example 8.8

Two 30A thyristors are connected in parallel to switch a load current of 50A.
The on-state characteristic of T_1 is $V_{T1(on)} = 1.2 + 56 \times 10^{-3} I_1$ (V), and for T_2 is $V_{T2(on)} = 1.1 + 40 \times 10^{-3} I_2$ (V).

(a) Calculate the current through each thyristor.

(b) A 0.1Ω series resistor is now connected in series with each thyristor to force more equal current sharing. Determine the new thyristor currents.

Solution

$I_2 = 50 - I_1$

(a) The on-state voltage drop across each thyristor must be the same, i.e.

$$1.2 + I_1 (56 \times 10^{-3}) = 1.1 + (50 - I_1)(40 \times 10^{-3})$$

$$0.1 = 2 - (40 + 56) \times 10^{-3} I_1$$

$$I_1 = 20.8 \,\text{A}$$

$$I_2 = 50 - I_1 = 50 - 20.8 = 29.2 \,\text{A}$$

(b) With series resistance of $0.1\,\Omega$,

$$1.2 + I_1 (56 + 100) \times 10^{-3} = 1.1 + (50 - I_1)(40 + 100) \times 10^{-3}$$

$$0.1 = 7 - 296 \times 10^{-3} I_1$$

$$I_1 = 23.3 \,\text{A}$$

$$I_2 = 50 - I_1 = 50 - 23.3 = 26.7 \,\text{A}$$

This represents a much better current sharing at the expense of increased forward voltage drop, and power loss, i.e. voltage across parallel combination without series resistors is

$$1.2 + 20.8 \times 56 \times 10^{-3} = 2.36 \,\text{V}$$

Voltage across parallel combination with series resistors is

$$1.2 + 23.3 \times 156 \times 10^{-3} = 4.83 \,\text{V}$$

Dynamic equalization can be achieved using coupled coils, as shown in Fig. 8.23.

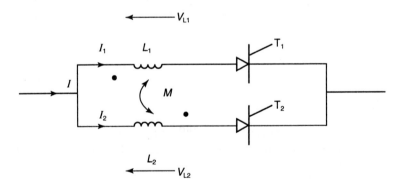

Figure 8.23

$$V_{L1} = L_1 \, \delta i_1/\delta t - M \delta i_2/\delta t$$

$$V_{L2} = L_2 \, \delta i_2/\delta t - M \delta i_1/\delta t$$

Assume $L_1 = L_2 = L$, and perfect coupling, giving $M = \sqrt{L_1 L_2} = \sqrt{L^2} = L$; then

$$V_{L1} = L \, (\delta i_1/\delta t - \delta i_2/\delta t)$$

$$V_{L2} = L \, (\delta i_2/\delta t - \delta i_1/\delta t)$$

If $\delta i_1/\delta t$ is changing at a faster rate than $\delta i_2/\delta t$, then V_{L1} acts to reduce i_1, and V_{L2} reverses its polarity, acting to increase i_2. Hence there is the tendency to dynamically balance the transient currents.

Gating

With series connected switches, the slowest to turn on will momentarily have the full voltage across it. During turn-off the fastest will have the full voltage across it.

With parallel connected switches, the first to turn on will momentarily carry the full current. At turn-off, the last to turn off will have the full current through it.

It is obviously desireable to turn on and turn off all the switches simultaneously. The gate-cathode circuits will not be identical, and to compensate for this a series resistance can be connected in the gate circuit of each switch. This will have the effect of reducing the spread of the gate currents.

Isolation of the gate circuits using pulse transformers or opto-isolators is necessary for series connected switches due to the very different cathode voltages. A simple gate circuit for series connected switches is shown in Fig. 8.24, and for parallel switches in Fig. 8.25.

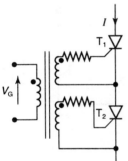

Figure 8.24   Figure 8.25

8.7 THE IGBT

The insulated gate bipolar transistor (IGBT) is effectively a Mosfet cascaded with a BJT. A simplified view of the semiconductor arrangement is shown in Fig. 8.26.

With gate and emitter at the same polarity and the collector positive, junction 2 is reverse-biased and no current flows from emitter to collector. With the gate positive with respect to the emitter and greater than the threshold voltage, the Mosfet channel is formed for current flow. This current is the base current for a

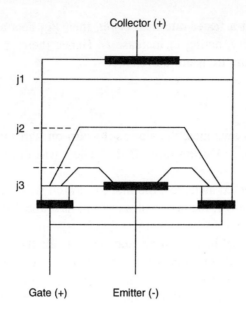

Figure 8.26

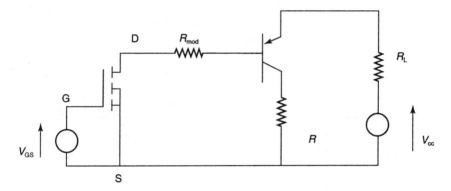

Figure 8.27

pnp transistor, allowing current to flow from emitter to collector, turning the switch on. A simplified equivalent circuit of the IGBT is given in Fig. 8.27.

The circuit symbol for the IGBT is usually shown as one of the alternatives given in Figs 8.28(a) and (b). There is no built-in parasitic diode in the IGBT for reverse breakdown. A separate diode is required.

The IGBT combines the easy gating requirements of the Mosfet with its high input impedence, and the power handling capability of the BJT. A relatively high gate-emitter voltage of 15 V or so will result in a forward collector-emitter voltage drop of around 3 V. This forward voltage drop will be higher for higher frequencies.

1200 V IGBTs are available with current ratings of 10–400 A. These will switch in about 1 μs with forward voltage drop between 2.5 and 5 V. A 500 V, 5 A IGBT might typically have a forward voltage drop of 2 V with a switching time of 250 ns.

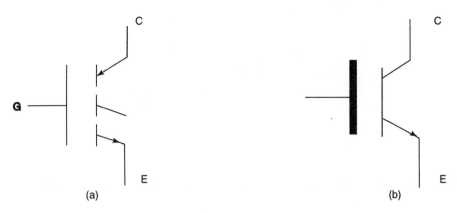

Figure 8.28

8.8   GATE AND BASE DRIVE ISOLATION

In circuits where a number of devices are to be switched on and off by a drive circuit with a number of outputs, all with a common ground, isolation is necessary to avoid shorting out part of the circuit. For example, in the full-wave single-phase bridge shown in Fig. 3.8, two of the four switches, T_3 and T_4, have a common cathode and these could be driven without the necessity for gate pulse isolation. However, the other two switches, T_1 and T_2, do not have a common cathode. Diagonally opposite thyristors are switched on together and this would result in a momentary short of the supply, as shown by the dotted line in Fig. 8.29, for the positive half-cycle with T_1 and T_3 conducting. In the case of Mosfets, the short would exist for the duration of the on-time.

The problem is overcome by isolation of the drive circuits using pulse transformers or opto-isolators. Figure 8.30 shows transformer isolation of the thyristor bridge. Where high frequency gate pulses are used, the isolation transformers are very small with very few turns.

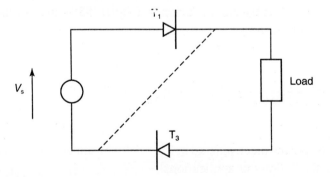

Figure 8.29

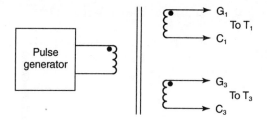

Figure 8.30

With Mosfets and IGBTs, optical isolation is often used. A typical arrangement is shown in Fig. 8.31.

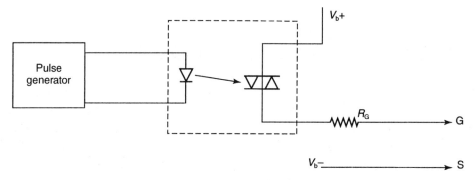

Figure 8.31

The opto-isolator shown in the dotted box in Fig. 8.31 can have as the switching device a Triac, thyristor, transistor or Darlington transistor. Manufacturers such as IR, Harris and Motorola supply these. Four opto-isolators are available in a single integrated circuit module. Complete drive circuits with overload protection and current and voltage sensing, the so-called 'smart ICs', are also available.

The drive circuits can of course be built up using discrete components. The pulse generator could be based on a Schmitt-Trigger, 555-timer, or the output of a logic circuit via a buffer.

8.9 SELF-ASSESSMENT TEST

1 What type of power electronic switch would you select, and in what configuration, for the following applications:

(a) variable d.c. voltage to a load from a fixed d.c. voltage source;

(b) variable d.c. voltage to a load from a.c. mains;

(c) variable a.c. voltage to a load from a.c. mains;

(d) variable a.c. frequency to a load from a d.c. voltage source?

2 What would be the disadvantage of using a thyristor instead of a Mosfet in a d.c. chopper circuit?

3 State the conditions required for the successful turn-on of a thyristor.

4 Explain the essential differences between the turn-on, or gating, requirements of thyristors and Mosfets.

5 Sketch typical off-state characteristics of two series connected thyristors and show how they share the voltage. What methods can be used to force more equal voltage sharing?

6 Sketch typical on-state characteristics of two parallel connected thyristors, and show how they share the current. What methods can be used to force more equal current sharing?

8.10 PROBLEMS

1 The on-state characteristic of a thyristor to be used in a d.c. chopper circuit is given by $V_{T(on)} = 1.1 + 0.02\ I$ (V), where I is the anode current in amperes. Calculate the average power dissipation in the thyristor when the anode current is 100 A and the chopper duty-cycle is 0.5.

2 In the gate pulse firing circuit shown in Fig. 8.6, the gate pulse magnitude is 15 V. A 1:1 isolating transformer is used. The limiting resistors are $R_1 = 15\Omega$ and $R_2 = 30\Omega$. If the linearized thyristor gate characteristic has a resistance of 20Ω, determine the gate current and voltage.

3 A thyristor has a latching current of 20 mA. It supplies current to a coil of 100Ω resistance and 1.0 H inductance from a 24 V d.c. supply. Calculate the value of the minimum duration of the gate pulse to ensure that the thyristor remains on when the gate pulse is removed.

4 A 2N6755 Harris n-channel Mosfet has a d.c. supply voltage of 60 V, and a load resistance of 6Ω. With a gate-source voltage of 10 V, the on-state resistance is 0.2Ω.

(a) Assuming that the Mosfet is acting as a simple on-off switch, determine the load current and efficiency.

(b) The Mosfet above is now used in a 20 kHz chopper circuit. Assuming the parameters remain unchanged, calculate the load power and efficiency at a chopper duty cycle of 0.4.

5 A resistive load is to be supplied at 3 kV and 1100 A. The thyristors available are rated at 800 V, 200 A. Calculate the number of series and parallel connected thyristors required, allowing one extra series and parallel thyristor as a safety factor.

Determine suitable values for the series static and dynamic equalizing components, when the maximum leakage current is 6 mA, the maximum difference in recovery charge is $20\,\mu C$, and the maximum rate of change of voltage is $200\,V/\mu s$.

Answers to self-assessment tests and problems

Chapter 2

Self-assessment test

Refer to text as outlined below for these answers.

1. Fig. 2.3, Fig. 2.5, equation (2.1). **2.** Fig. 2.11. **3.** Fig. 2.15, Fig. 2.16. **4.** Equations (2.8)–(2.13).

Problems

1. (a) $V_{av} = V_b \, t_n f = 60 \times 0.2 \times 10^{-3} \times 10^3 = 12\,V$
$I_{av} = V_{av}/R = 12 \,/\, 12 = 1\,A$
$V_{rms} = V_b \sqrt{t_n f} = 60 \sqrt{0.2} = 26.83\,V$
$I_{rms} = V_{rms}/R = 26.83/12 = 2.24\,A$
$P = I_{rms}^2 \, R = (2.24)^2 \times 12 = 60\,W$

   (b) $V_{av} = V_b t_n f = 60 \times 0.6 \times 10^{-3} \times 10^3 = 36\,V$
$I_{av} = V_{av} \,/\, R = 36 \,/\, 12 = 3\,A$
$V_{rms} = V_b \sqrt{t_n f} = 60 \sqrt{0.6} = 46.5\,V$
$I_{rms} = V_{rms}/R = 46.5/12 = 3.87\,A$
$P = I_{rms}^2 \, R = (3.87)^2 \times 12 = 180\,W$

2. (a) $V_{av} = V_b \, t_n f = 24 \times 8 \times 10^{-3} \times 100 = 19.2\,V$

During turn-on,

$i(t) = (V_b/R) \, (1 - \exp(-Rt/L)) + I_0 \exp(-Rt/L)$
$V_b/R = 24/1 = 24$
$Rt/L = 1 \times 8 \times 10^{-3} \,/\, 10^{-2} = 0.8$

At end of turn-on period, $I(t) = I_1$. Therefore

$I_1 = 24(\, 1 - \exp(-0.8)) + I_0 \exp(-0.8)$
$I_1 = 13.22 - 0.449 \, I_0$ (1)

During turn-off, $i(t) = I_1 \exp(-Rt/L)$.

$t_n = 8\,ms, \; T = 1/f = 1/\,100 = 10\,ms, \; t_f = T - t_n = 2\,ms$
$Rt/L = 1 \times 2 \times 10^{-3} \,/\, 10^{-2} = 0.2$

At the end of turn-off period, $I(t) = I_0$. Therefore

$I_0 = I_1 \exp(-0.2) = 0.819 \, I_1$ (2)

Equation (2) substituted in (1) gives

$I_1 = 13.22 - 0.449(0.819)I_1$

$$I_1 (1 - 0.368) = 13.22$$
$$I_1 = 20.9\,\text{A}$$
$$I_0 = 0.819\,I_1 = 17.1\,\text{A}$$

(b)   Follow the procedure above to obtain:

$$V_{av} = 4.8\,\text{V},\ I_1 = 6.88\,\text{A},\ I_0 = 3.09\,\text{A}$$

3. $R = R_a + R_f = (0.2 + 0.2) = 0.4\,\Omega$
$E = V - I_a\,R = 48 - (8 \times 0.4) = 44.8\,\text{V}$
$k_v = E/\omega\,I_a = 44.8/(3000 \times 2\pi/60)\,8 = 0.0178\ \text{V/A-rad/s}$

With $I_a = 80\,\text{A}$:

$V_{av} = Vt_n f = 48 \times 3.5 \times 10^{-3} \times 250 = 42\,\text{V}$
$E = V_{av} - I_a\,R = 42 - (80 \times 0.4) = 10\,\text{V}$
$\omega = E/k_v I_a = 10/0.0178 \times 80 = 7.022\ \text{rad/s}$
$N = \omega \times 60/2\pi = 67\ \text{rev/min}$
$T = k_v I_a{}^2 = 0.0178 \times 80^2 = 113.9\,\text{Nm}$

4. (a) $T = 5\,\text{Nm},\ N = 1500\,\text{rev/min},\ k_v = 10^{-2}\ \text{V/A-rad/s}$
$T = k_v I_a{}^2 \ \therefore\ I_a = \sqrt{T/k_v} = \sqrt{5/10^{-2}} = 22.36\,\text{A}$
$\omega = N \times 2\pi/60 = 1500 \times 2\,\pi/60 = 157.1\ \text{rad/s}$
$E = \omega\,k_v I_a = 157.1 \times 10^{-2} \times 22.36 = 35.1\,\text{V}$
$V_{av} = E + I_a R = 35.1 + (22.36 \times 0.1) = 37.4\,\text{V}$
$V_{av} = Vt_n f \ \therefore\ t_n = V_{av}\,/\,Vf = 37.4/\,96 \times 125 = 3.11\,\text{ms}$

(b) $T = 20\,\text{Nm},\ N = 1500\,\text{rev/min}$
$I_a = \sqrt{T/k_v} = \sqrt{20/10^{-2}} = 44.7\,\text{A}$
$E = \omega\,k_v I_a = 157.1 \times 10^{-2} \times 44.7 = 70.3\,\text{V}$
$V_{av} = E + I_a R = 70.3 + (44.7 \times 0.1) = 74.8\,\text{V}$
$V_{av} = Vt_n f \ \therefore\ t_n = V_{av}/Vf = 74.8/96 \times 125 = 6.23\,\text{ms}$

5. Load voltage, $V_0 = V_b (1 - D)$. Therefore

$(1 - D) = V_b\,/\,V_0 = 12/48 = 0.25$
Hence duty cycle, $D = 0.75$
Load current, $I_L = V_0/R_L = 48/12 = 4\,\text{A}$
Battery current, $I_b = I_L/(1 - D) = 4/0.25 = 16\,\text{A}$
$T = 1/f = 1/10^5 = 10^{-5}\,\text{s}$
$\Delta I = V_b t_n/L = V_b\,DT/L = 12 \times 0.75 \times 10^{-5}\,/\,10^{-5} = 9\,\text{A}$
Maximum battery current, $I_1 = I_b + \Delta I/2 = 16 + 4.5 = 20.5\,\text{A}$
Minimum battery current, $I_0 = I_b - \Delta I/2 = 16 - 4.5 = 11.5\,\text{A}$

Chapter 3

Self-assessment test

1. (a) Half-controlled bridge; (b) fully-controlled bridge; (c) inverse-parallel fully controlled bridge.

2. See Fig. 3.11.

3. See Fig. 3.12.

4. $V_{av} - (2E_m/\pi)\cos a$ (V); $E = k_v\,\omega$ (V); $I_{av} = (V_{av} - E)/R_a$ (A); $T = k_v\,I_{av}$ (Nm).

Average armature voltage must exceed generated voltage in order for the thyristor to turn on.

5. $V_{av} = (E_m/\pi)(1 + \cos a)$ (V); $E = k_v \, \omega$ (V); $I_{av} = (V_{av} - E)/R_a$ (A); $T = k_v \, I_{av}$ (Nm). Firing angle delay range is limited as in the case of the fully controlled bridge.

6. $E_m = \sqrt{2} \times 250 = 354\,V$
$P = 2$, $V_{av} = (2E_m/\pi)\cos a = (2 \times 354/\pi)\cos 0° = 225\,V$
$P = 3$, $V_{av} = (3\sqrt{3}\,E_{pm}/\pi)\cos a = (5.2 \times 354/2\pi)\cos 0° = 293\,V$
$P = 6$, $V_{av} = (3\sqrt{3}\,E_{pm}/\pi)\cos a = (5.2 \times 354/\pi)\cos 0° = 586\,V$
$P = 12$, $v_{av} = (6\sqrt{3}\,E_{pm}/\pi)\cos a = (10.4 \times 354/\pi)\cos 0° = 1172\,V$

7. A three-phase, full-wave converter with an armature contactor, or two anti-parallel bridges. Assume the use of an armature contactor. Firing angle is increased to bring armature current to zero, contactor is operated and firing angle increased, converter voltage is reversed, armature voltage exceeds converter voltage and energy is returned to the supply. Speed falls to zero as motor brakes. Firing angle is now reduced to below 90° for rectifier operation, and speed builds up in the reverse direction.

Problems

1. For circuit see Fig. 3.6; for waveform see Fig. 3.11.
   (a) $V_{rms} = E_s \sqrt{1 - (a/\pi) + (\sin 2a)/2\pi}$
   $= 110 \sqrt{1 - (45°/180°) + (\sin 90°)/2\pi} = 104.9\,V$
   $I_{rms} = V_{rms}/R = 104.9/100 = 1.049\,A$
   $P = I_{rms}^2 \times R = 1.049^2 \times 100 = 110\,W$

   (b) Follow procedure above with $a = 135°$ to obtain $P = 11\,W$.

2. (a) *Resistive load.* Use equations (3.5) to (3.9) with power, $P = I_{rms}^2 \times R$, and power-factor, $\cos\phi = P/E_{rms} \times I_{rms}$. Solutions tabulated below:

$a°$	V_{av} (V)	I_{av} (A)	V_{rms} (V)	I_{rms} (A)	P (W)	$\cos\phi$
30	202	0.81	237	0.95	225	0.99
60	162	0.65	227	0.91	207	0.95

   (b) *Highly inductive load.* Use procedures outlined from equations (3.11) to (3.12). Solutions tabulated below:

$a°$	V_{av} (V)	I_{av} (A)	V_{rms} (V)	I_{rms} (A)	P (W)	$\cos\phi$
30	132	0.53	250	0.53	70	0.53
60	76.4	0.31	250	0.31	24	0.33

3. (a) The resistive load values are the same as for the fully controlled resistive load case in 2(a) above.

   (b) *Highly inductive load.* Apply the expressions used in Example 3.4 (p. 38) to obtain the answers given below:

$a°$	V_{av} (V)	I_{av} (A)	V_{rms} (V)	I_{rms} (A)	P (W)	$\cos\phi$
30	202	0.81	237	0.74	164	0.92
60	162	0.65	227	0.53	106	0.83

4. (a) $a = 30°$:

   $V_{av} = (2E_m/\pi) \cos a = (2\sqrt{2} \times 250/\pi)\cos 30° = 195\,V$

$E = k_v \, \omega = 0.9 \times 1200 \times 2\pi/60 = 113 \, \text{V}$
$I_{av} = (V_{av} - E)/R_a = (195 - 113)/0.75 = 109 \, \text{A}$
$T = k_v \, I_{av} = 0.9 \times 109 = 98 \, \text{Nm}$

(b) $a = 70°$; $E = 113.1 \, \text{V}$, $V_{av} = 117 \, \text{V}$, $I_{av} = 5.2 \, \text{A}$, $T = 4.7 \, \text{Nm}$.

5. For the half-controlled bridge, $V_{av} = (\sqrt{2} \, E_s/\pi)(1 + \cos a)$, then follow the pattern of calculations outlined in 4(a) above, to obtain:

(a) $a = 30°$; $E = 113.1 \, \text{V}$, $V_{av} = 210 \, \text{V}$, $I_{av} = 59.1 \, \text{A}$, $T = 53.2 \, \text{Nm}$;

(b) $a = 70°$; $E = 113.1 \, \text{V}$, $V_{av} = 151 \, \text{V}$, $I_{av} = 50.5 \, \text{A}$, $T = 45.5 \, \text{Nm}$.

6. The armature voltage for each bridge is given by

(a) $V_{av} = (2E_m/\pi)\cos a$

(b) $V_{av} = (E_m/\pi)\cos a$

(c) $V_{av} = (3\sqrt{3} \, E_{pm}/2\pi)\cos a$

(d) $V_{av} = (3\sqrt{3} \, E_{pm}/\pi)\cos a$.

Again calculations follow the pattern of problem 4(a) above to yield the answers below:

(a) $I_{av} = 12.4 \, \text{A}$, $T = 9.9 \, \text{Nm}$

(b) $I_{av} = 43 \, \text{A}$, $T = 34.4 \, \text{Nm}$

(c) $I_{av} = 56.1 \, \text{A}$, $T = 44.9 \, \text{Nm}$

(d) $I_{av} = 246 \, \text{A}$, $T = 197 \, \text{Nm}$

7. $V_{av} = (3\sqrt{3} \, E_{pm}/2\pi)\cos a = (3\sqrt{2} \times 415/2\pi)\cos a = 280.1 \cos a$
$I_{av} = T/k_v = 400/0.8 = 500 \, \text{A}$
$E = k_v \, \omega = 0.8 \times 1000 \times 2\pi/60 = 83.8 \, \text{V}$
$V_{av} = E + I_{av}R$, i.e. $280.1 \cos a = 83.8 + (500 \times 0.02) = 93.8$
$\therefore \cos a = 93.8/280.1 = 0.335$
$a = 70.4°$

Chapter 4

Self-assessment test

1. Inverter frequency is controlled by the rate at which the semiconductor switches are turned on and off, i.e. by the periodic time of the base drive pulses. Output voltage is controlled by the on-time of each switch, i.e. by the width of the switch on-time compared with the half-periodic time.

$$V_{rms} = \sqrt{(2/T) \int_0^{t_{on}} (V_b)^2 \, \delta t} = V_b \sqrt{(2/T) \, t_{on}}$$

Maximum time-on, $t_{on} = T/2$.

2. See Figs 4.4 and 4.5.

$V_{c1(max)} = V_b/(1 + \exp(-T/2CR))$
$V_{c1(min)} = V_{c1(max)} \exp(-T/2CR)$

3. See Figs 4.6 and 4.7.
$I_0 = V_b/4fL = V_b \, T/4L$

4. See Fig. 4.11.

Q_1 on: $i(t) = (V_b/R) (1 - \exp(-Rt/L)) - I_0 \exp(-Rt/L)$
Q_2 on: $i(t) = I_0 \exp(-Rt/L) - (V_b/R)(1 - \exp(-Rt/2L))$
$I_0 = (V_b/R)((1 - \exp(-Rt/2L))/(1 + \exp(-Rt/L))$

Problems

1. $T = 1/f = 1/1000 = 1\,\mathrm{ms}$
$CR = 100 \times 10^{-6} \times 5 = 0.5\,\mathrm{ms}$
$V_{c1(max)} = V_b/(1 + \exp(-T/2CR)) = 50/(1 - \exp(-1)) = 36.6\,\mathrm{V}$
$V_{c1(min)} = V_b \exp(-T/2CR) = 50 \exp(-1) = 18.4\,\mathrm{V}$

Load current waveform:

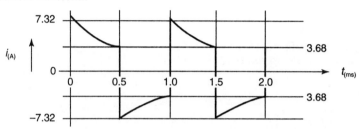

Figure P4/1

2. $T = 1\,\mathrm{ms}$
$I_0 = V_b T/4L = 50 \times 10^{-3}/4 \times 10^{-3} = 12.5\,\mathrm{A}$

Load current waveform:

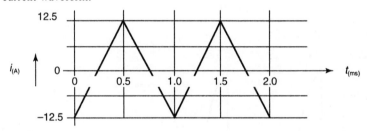

Figure P4/2

3. $T = 1\,\mathrm{ms}$
$L = 1\,\mathrm{mH}, R = 5\,\Omega$

From equation (4.6),

$I_0 = (V_b/R)((1 - \exp(-RT/2L))/(1 + \exp(-RT/2L)))$
$= (50/5)((1 - \exp(-2.5))/(1 + \exp(-2.5))) = 8.48\,\mathrm{A}$

Load current waveform:

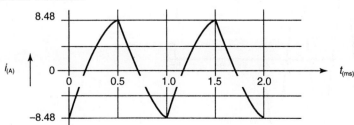

Figure P4/3

4. $\omega_0^2 = 1/LC$

$\therefore L = 1/\omega_0^2 C = 1/(2\pi \times 10^3)^2 \times 10^{-4} = 253\,\mu H$

5. $i_C = I_m\sin \omega t$

At $t = t'$, $20 = 30 \sin \omega_0 t$. Therefore $0.667 = \sin \omega_0 t'$.

$\omega_0 t' = \sin^{-1} 0.667 = 41.84° = 0.73$ rad

$\omega_0 t_{off} = 180 - 2\omega_0 t' = 96.32° = 1.682$ rad

$\omega_0 = 1.682/t_{off} = 1.682/40 \times 10^{-6} = 42050$ rad/s

$\therefore f_0 = 6692$ Hz

Equating energy storage: $0.5\,LI^2 = 0.5\,CV^2$. From this

$L = CV^2/I^2 = C\,(96)^2/(30)^2 = 10.24\,C$

Now

$\omega_0^2 = 1/LC = 1/10.24\,C^2$

$C = 1/(\sqrt{10.24} \times 42050) = 7.43\,\mu F$

$L = 10.24 \times 7.43 \times 10^{-6} = 76\,\mu H$

Chapter 5

Self-assessment test

1. Phase control driver requires a zero crossing detector and time delay circuit to vary the firing angle delay. Burst-firing driver also requires a zero crossing detector in combination with a pulse-width modulator and logic circuitry to control the number of cycles passed to, and the number of cycles blocked from, the load.

2. (a) $V_{rms} = (E_s) \sqrt{(1 - \alpha/\pi + (\sin 2\alpha)(/2\pi)}$

   (b) $V_{rms} = E_s \sqrt{t_n/(t_n + t_f)}$

   (c)

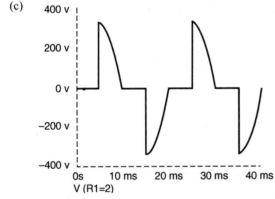

   (d)

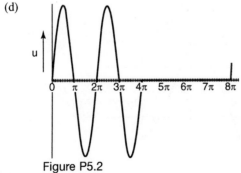

   Figure P5.2

3. $i = (E_m/|Z|) \{ \sin (\omega t + \alpha - \phi) + \sin (\phi - \alpha) \exp(-Rt/L)\}$

(a) $\alpha = \phi$, $i = (E_m/|Z|) (\sin (\omega t)$; waveform as Fig. 5.4.

(b) $\alpha > \phi$; waveform as Fig. 5.5.

4. (a) See Fig. 5.12; (b) see Fig. 5.13.

5. The advantage of burst-firing is that the load voltage waveform has a much lower harmonic content and therefore less generation of interference. The disadvantage is that the method is only suitable for loads with long thermal time constants.

Problems

1. (a) $V_{rms} = 246\,V$, $I_{rms} = 1.971\,A$, $P = 486\,W$, $\cos\phi = 0.985$

(b) $V_{rms} = 43$, $I_{rms} = 0.341\,A$, $P = 14.5\,W$, $\cos\phi = 0.17$

2. (a) $V_{rms} = 53.7\,V$, $I_{rms} = 1.074\,A$, $P = 57.7\,W$, $\cos\phi = 0.448$

(b) $V_{rms} = 107.3$, $I_{rms} = 0.215\,A$, it*P $= 230\,W$, $\cos\phi = 0.89$

3. 2083 W in each case. Burst-firing produces fewer harmonics and is the preferred method. Phase control would give a finer adjustment of load power.

4. $\beta = 171.7°$.

5. $I_{av} = 2.86\,A$, $T_{av} = 1.72\,Nm$.

Chapter 6

Self-assessment test

1. See Fig. 6.8.

2. (a) $n_s = f/p = 50/6 = 8.33\,rev/s$. $N_s = 60n_s = 500\,rev/min$.

(b) $n_s = f/p = 50/3 = 16.67\,rev/s$. $N_s = 60n_s = 1000\,rev/min$.

(c) $n_s = f/p = 50/1 = 50\,rev/s$. $N_s = 60n_s = 3000\,rev/min$.

3. $n_r = n_s (1 - s)$.

(a) $n_r = (50/1) (1 - 0.02) = 49\,rev/s$, $N_r = 60 \ n_s = 2940\,rev/min$.

(b) $n_r = (50/1)(1 - 0.05) = 47.5\,rev/s$, $N_r = 60n_s = 2850\,rev/min$.

(c) $n_r = (50/1) (1 - 0.03) = 48.5\,rev/s$, $N_r = 60n_s = 2910\,rev/min$.

At 80 Hz, all rotor speeds are increased by a factor (8/5).

4. $\Phi = E_s/kf \approx V_s/kf$.

(a) If V_s is constant and frequency is decreased then flux, Φ, will tend to saturate, distorting current waveforms and altering motor parameters. If V_s is constant and frequency is increased, flux will be reduced, motor reactances will be reduced, and current will increase beyond rated value.

(b) The stator voltage drop at low frequencies becomes more significant compared to the reduced stator applied voltage. A boost in applied voltage is necessary to hold the flux constant.

5.

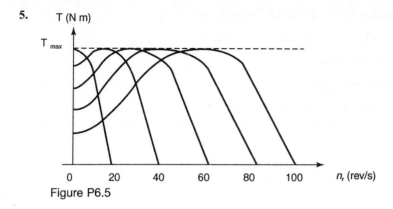

Figure P6.5

Problems

1. $N_r = 380$ rev/min, $I_s = 31.31$ $\underline{/- 47.2°}$ (A), $I_r = 21.5$ $\underline{/-3.9°}$ (A)
   $T = 132.4$ N m.

2. Line voltage $= 495$ V, $N_r = 1151$ rev/min.

Chapter 7

Self-assessment test

1. Smaller size of inductors and transformers. Higher efficiency.

2. (a) Isolation between input and output is achieved.

   (b) To prevent the build-up of d.c. magnetization of the transformer saturating the core.

3. (a) See Fig. 7.8.

   (b) $V_0 = V_1 (N_s/N_p) (D/1- D)$.

   (c) See Figs 7.1 and 7.2, and the following explanation.

4. Push–pull, $V_0 = V_1 (N_s/N_p) 2 D$; half-bridge, $V_0 = V_1 (N_s/ N_p) D$; full-bridge, $V_0 = V_1 (N_s/N_p) 2D$

5. See Fig. 7.14.

6. The forward converter transfers power to the load during the on-time of the power electronic switch. Output voltage is less than the input voltage.
   The flyback converter transfers power to the load during the off-time of the power electronic switch. Output voltage is more than the input voltage.

Problems

1. (a) $P = 125$ W, $I_L = 25$ A, $D = 0.4$.

   (b) $t_n = 10$ μs, $t_f = 15$ μs, $(I_1 - I_0) = 2.5A$,
   $i_{L(min)} = 23.75$ A, current is continuous.

2. (a) $D = 0.67$, $R = 4.61$ Ω.

   (b) $I_L = 10.4$ A, $i_{L(max)} = 14.42$ A, $i_{L(min)} = 6.38$ A.

3. (a) $L = 30$ μH, $C = 470$ μF.

(b) $I_L = 20\,\text{A}$, $i_{L(\text{max})} = 45\,\text{A}$, $i_{L(\text{min})} = 5\,\text{A}$.

(c) $V_1 = 150\,\text{V}$, $\Delta V_0/V_0 = 1.3\%$.

Chapter 8

Self-assessment test

1. (a) A Mosfet if the rating is available, otherwise a BJT. The power swtich would operate as a chopper.

   (b) Thyristors in a full-wave, half-controlled or fully controlled circuit depending on circuit requirements.

   (c) A Triac as an a.c. controller.

   (d) Mosfets in an inverter bridge configuration.

2. Thyristors require forced commutation circuits when the supply is d.c.

3. Manufacturer's data sheets must be consulted to ascertain anode and gate rated values. Anode polarity must be positive with respect to the cathode. Gate pulse duration must be long enough for the anode current to exceed the latching current. The higher the gate current, the faster the turn-on. Gate power must not exceed the rated maximum value.

4. Thyristor gates are current-driven, Mosfet gates are voltage driven. Once the thyristor is on, the gate pulse can be removed. The Mosfet requires a continuous gate voltage for the duration of the on-time; however, apart from the initial charge at swtich-on, the gate current is negligible.

5. See Fig. 8.18. Parallel resistors for static voltage equalization. Parallel R–C network for dynamic voltage equalization.

6. See Fig. 8.21. Series resistors for static current equalization. Coupled coils for dynamic current equalization.

Problems

1. $V_{\text{Ton}} = 1.1 + 0.03\,I = 1.1 + 0.02 \times 100 = 3.1\,\text{V}$
   $P_{\text{max}} = V_{\text{T(on)}} \times I = 3.1 \times 100 = 310\,\text{W}$
   $P_{\text{av}} = P_{\text{max}}\,t_n/T = 310 \times 0.5 = 155\,\text{W}$

2. The Thevinin equivalent circuit of the gate firing circuit is shown below:

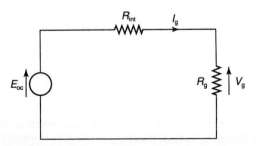

$E_{\text{oc}} = V_G \times R_2/(R_1 + R_2) = 15 \times 30/(15 + 30) = 10\,\text{V}$
$R_{\text{int}} = R_1\,R_2/(R_1 + R_2) = 15 \times 30/(15 + 30) = 10\,\Omega$
$I_G = E_{\text{oc}}/(R_{\text{int}} + R_g) = 10/(10 + 20) = 0.33\,\text{A}$
$V_g = I_g\,R_g = 0.33 \times 20 = 6.7\,\text{V}$

3. The anode current will increase exponentially, i.e.

$$i = I_{max} (1 - \exp(-Rt/L)$$

where $I_{max} = V/R = 24/100 = 0.24\,\text{A}$.

$$R/L = 100/1 = 100$$

To reach latching current

$i = I_{max} (1 - \exp(-Rt/L))$ or $0.02 = 0.24(1 - \exp(-100t))$
$0.0833 = 1 - \exp(-100t)$
$\exp(-100t) = 1 - 0.0833 = 0.9167$
$-100t = I_n (0.9167) = -0.087$
$t = 870\,\mu s$

4. (a) $I_D = V_{DD}/(R_{DS(on)} + R_L) = 60/(6 + 0.2) = 9.68\,\text{A}$
Load power $= (I_D)^2\, R_L = (9.68)^2 \times 6 = 562\,\text{W}$
Power dissipated in Mosfet $= (I_D)^2\, R_{DS(on)} = (9.68)^2 \times 0.2 = 18.7\,\text{W}$
Efficiency, $\zeta = 562/(562 + 18.7) = 97\%$

(b) From equation (2.3), $V_{rms} = V_{DD}\sqrt{t_n \times f}$.
$t_n = DT = D/f = 0.4/(20 \times 10^3) = 20\,\mu s$
$V_{rms} = V_{DD}\sqrt{t_n \times f} = 60\sqrt{20 \times 10^{-6} \times 20 \times 10^3} = 37.9\,\text{V}$

$$I_{rms} = V_{rms}/R_L = 37.9/(6 + 0.2) = 6.12\,\text{A}$$

Load power $= (I_{rms})^2\, R_L = (6.12)^2 \times 6 = 225\,\text{W}$

Input power $= (6.12)^2 \times (6 + 0.2) = 232.2\,\text{W}$
Efficiency $\zeta = 225/(232.2) = 97\%$

Alternatively, using an averaging technique load power,

$$P_{av} = P_{max}\, t_n f = 562 \times 0.4 = 225\,\text{W}$$

6. For series connection

$$n_s = (V_s/V_D) + 1 = (3000/800) + 1 = 5$$

For parallel connection

$$n_p = (I_s/I_p) + 1 = (1100/200) + 1 = 7$$

From equation (8.6)

$R_1 = (nV_D - V_s)/(n - 1)I_0$
$\quad = ((5 \times 800) - 3000)/(5 - 1) \times 6 \times 10^{-3} = 42\,\text{k}\Omega$

From equation (8.7)
$C = (n-1)\Delta Q/nV_D - V_s$
$\quad = (5 - 1) \times 20 \times 10^{-6}/(5 \times 800) - 3000 = 0.08\,\mu F$
Choose $C = 0.1\,\mu F$.

A value of $R_2 = 20\,\Omega$ will give an initial maximum discharge current of about $800/20 = 40\,\text{A}$, with a maximum rate of change of voltage $(\delta v/\delta t)_{max} = -V/CR_2 = -800/0.1 \times 10^{-6} \times 20 = -400\,\text{V}/\mu s$.

References and bibliography

Bird, B.M., King, K.G., Pedder, D.A.G. 1993: *An introduction to power electronics*. Chichester: Wiley.

Bradley, D.A. 1987: *Power electronics*. Wokingham: Van Nostrand.

Hart, D.W. 1997: *Introduction to power electronics*. New York: Prentice-Hall.

Kusko, A. 1969: *Solid-state DC motor drives*. Cambridge, MA: MIT Press.

Lander, C.W. 1993: *Power electronics*. London: McGraw-Hill.

Larson, B. 1983: *Power control electronics*. New York: Prentice-Hall.

Murphy, J.M.D., Turnbull, F.G. 1989: *Power control of AC motors*. Oxford: Pergamon.

Pearman, R. 1980: *Power electronics solid state motor control*. New York: Prentice-Hall.

Ramshaw, R., Schurman, D. 1997: *PSPICE simulation of power electronic circuits*. London: Chapman & Hall.

Ramshaw, R.S. 1993: *Power electronic semiconductor switches*. London: Chapman & Hall.

Rashid, M.H. 1993: *Power electronics*. New York: Prentice-Hall.

Vithayathil, J. 1995: *Power electronics*. New York: McGraw-Hill.

Williams, B.W. 1992: *Power electronics*. Basingstoke: Mcmillan.

Index